CU01572823

Fundamentals of Agronomy

THE AUTHORS

Dr Rajeev Singh presently working as Senior Scientist and Head at Krishi Vigyan Kendra Gaya-I which comes under administrative control of Bihar Agricultural University, Sabour, Bhagalpur. He did his Master Degree in Agronomy from and Ph.D. from Narendra Dev University of Agriculture and technology Faizabad in 2005. Dr Singh is working on various crops like Paddy, Lentil, Wheat, chick pea, Mustard and other major field crops and focusing on resource conservation technologies, nutrient management, weed management and other agronomic practices. He has published about 19 research papers in National/ International journals. He has also published 3 book, 2 book chapters and many articles. He has also handled Projects of IRRI, ICARDA, NICRA, Biotech Kisan Hub, Seed hub and Climate resilient program. He has also got Best extension Scientist award, Best senior Scientist award and fellowship award from different National societies.

Dr Vivek Yadav presently working as Assistant Agronomist at Rice Research Station, Nagina, Bijnor which comes under administrative control of Sardar Vallabhbhai Patel University of Ag & Technology, Meerut. He did his Master Degree in Agronomy from Agra University, Agra and PhD From Narendra Dev University of Agriculture and technology Faizabad in 2005. Dr Yadav is working on rice agronomy and focusing on resource conservation technologies of rice. He has published about 32 research papers in National/ International journals. He has also published 3 book, 7 book chapters and many articles. He has also been the potential reviewer of many journals and handled Projects of L G Life Science, Japan, Sumitomo Chemicals Pvt Ltd, Japan and Utter Pradesh council of Agricultural Research, Lucknow. He has also got Young Scientist Award, Excellence in Research Award and Scientist of the Year award from different National societies.

Shri Ajai Kumar Yadav presently working as Director, Regional Fodder Station, Chennai, Ministry of Fisheries, Animal Husbandry & Dairying, Department of Animal Husbandry & Dairying, Govt. of India. He did his Master Degree in Agronomy from Narendra Dev University of Agriculture and Technology, Faizabad. Shri Yadav is working on agronomy of fodder crops and quality fodder seed production with a vision and strategy to enhance livestock sector productivity by addressing need of feed and fodder. He has published about 18 research papers in National/ International journals. He has also published 3 Manuals/ book chapters and many Abstracts and articles.

Fundamentals of Agronomy

As per the 5th Deans Committee Report of ICAR

As a Core Subject in the Curriculum of B.Sc. (Agriculture) Course of the Agricultural Universities

— *Authors* —

Rajeev Singh

Senior Scientist and Head
Krishi Vigyan Kendra Gaya-I
Bihar Agricultural University, Sabour, Bhagalpur

Vivek Yadav

Assistant Agronomist
Rice Research station, Nagina, Bijnor

Ajai Kumar Yadav

Director
Regional Fodder Station, Chennai

2022

Daya Publishing House®
A Division of

Astral International Pvt. Ltd.
New Delhi – 110 002

Published by : **Daya Publishing House®**
A Division of
Astral International Pvt. Ltd.
– ISO 9001:2015 Certified Company –
4736/23, Ansari Road, Darya Ganj,
New Delhi-110 002
Ph. 011-43549197, 8130496929
E-mail: info@astralint.com
Website: www.astralint.com

Preface

Agronomy provides farmers with agricultural information about how to grow and care for plants and soils in certain environments. Factors such as climate, roots, moisture, weeds, pests, fungi, and erosion can all pose significant challenges when farmers attempt to produce a plentiful harvest. In order to discover ways of integrating crops into the environment in ways that will allow them to prosper, agronomists study these agricultural hurdles. Agronomists test soil and plant samples from all over the world to better understand how their properties and genetic make-up will interact with their surrounding environment. They then take this research to develop innovative farm practices and technologies that will boost crop yields and protect against weeds and pests. These findings give farmers across the globe crucial information and tips for not only producing crops but also for conserving our natural environment. Concerns of food and water security, air quality and climate change, soil loss and degradation, health and nutrition, and many others motivate agronomists to continually explore our agricultural resources.

In the earlier times, India was largely dependent upon food imports, but the successive story of the agriculture sector of Indian economy has made it self-sufficing in grain production. The country also has substantial reserves for the same. India depends heavily on the agriculture sector, especially on the food production unit after the 1960 crisis in food sector. Since then, India has put a lot of effort to be self-sufficient in the food production and this endeavour of India has led to the Green Revolution. The Green Revolution came into existence with the aim to improve the agriculture in India

This book has been designed to cover undergraduate syllabus of Fundamentals of Agronomy as per ICAR V Deans' Committee report. It covers the entire syllabus in one compact volume of 15 chapters. It covers introduction, seeds and sowing, tillage, crop density and geometry, manures and fertilizers, essential plant nutrients, irrigation management, crop rotation, allelopathy, herbicides, weed management, cropping systems, factors affecting crop production, plant ideotypes, adaptation and distribution of crops and harvesting and threshing. Related terms have been provided at the end for the ready reference of readers. The present book has been recommended as a core subject in the curriculum of B.Sc. (Agriculture) course of the agricultural universities. This book would be immensely useful to students, teachers, researchers, progressive farmers and extension workers working in the field of crop science, agronomy and agriculture.

We are grateful to all those persons as well as various books, manuals, periodicals, magazines, journals *etc.* that helped in the preparation of this book. In spite of the best efforts, it is possible that some errors may have occurred into the compilation and editing of the book. Further queries, constructive suggestions and criticisms for the improvement of the book are always welcome and shall be thankfully acknowledged.

Rajeev Singh

Vivek Yadav

Ajai Kumar Yadav

Contents

Chapter 1

Introduction

1. Agronomy

Agronomy is one of the subject of agriculture or agricultural science and the term agronomy has been derived from Greek language which includes *agros* meaning 'field' and *nomos*' meaning 'to manage'. Agronomy is the art of managing field, also the science of field crop production and management, and hence the science and economics of crop production by management of farm land.

Agronomy deals with the interaction of physical environment of soil and water with the crop in terms of the nutrients required by the crop plants of different species and when and why to supply the nutrients, growth and development of crops, effect of climate and other environmental factors on crop production, and the best way to control the potential competitors like weeds and other crop plants, insects/pests, and pathogens.

Therefore, Agronomy is the branch of agriculture science which covers principles and practices of soil, water, and crop management by providing favourable environment (external conditions and their effects in affecting the life and development of an organism) to the production and management of field crops for their higher productivity.

Agronomy among the different branches of agriculture occupies a prime position and regarded as a mother subject of different branches of agriculture as it is an integrated and applied aspect of different subjects of pure sciences. Agronomy has mainly three main branches *viz. crop science* mainly the field crops, *soil science*, and *environmental science* to deal with applied aspects. Agronomy has soil-crop-environment relationship as its central theme. This is because the soil is barren without crops and there is no existence of field crops without soil. Likewise is the role of environment (micro-climate) in enhancing the productivity of field crops.

The major aspects of the study of agronomy to enhance production per unit of land, time and input resources are:

> ☆ The yield variation with respect to cause and effect relationship,
>
> ☆ Internal and external factors with their interrelationships,
>
> ☆ Technologies for increasing the use-efficiency of inputs, and
>
> ☆ Evolving new technologies for better management practices of soil, water, plant nutrients, control of weeds, and also for crop plants.

In view of the above aspects of agronomy, the production of field crops on scientific lines mainly covers the crop improvement by adopting the improved agro-techniques in association of ameliorating environmental (agro-climate) conditions of the locality for entire duration of the crop.

2. Basic Principles of Agronomy

The principles of agronomy are based on fulfilling its aim to obtain maximum return from a crop per unit area for years together. This is possible to make improvement in genetic potential of field crops and to provide better environment by adopting the techniques for better management of soil, plants and environmentally controlled factors. The improvement in the genetic potential is the subject matter of plant geneticist and plant breeder by way of breeding methods and selection of superior germ plasm. The factors under the control of environment are drought, strong wind (gale), attack of insects/pests in particular year, *etc.* and these can be modified to some extent. Efforts can be made to mitigate the adverse effects of these adverse environmental factors to some extent by making arrangement for irrigation of field and by providing shelterbelts. However, the type of farming (specialized, diversified, mixed, and integrated) as well as the facilities available with the farmers (rainfed or *barani* farming, dry land agriculture, and irrigated farming) decides the principles of crop management:

> ☆ *Choice of crop* and the variety depending on the agro-climate of the area, fertility status of soil, location of land, season, and method of cultivation;
>
> ☆ Selecting *quality seed* of the crop and to maintain proper plant density by way of proper seed rate or by adopting the practices of gap filling or thinning;
>
> ☆ Adopting *multiple cropping* and also mixed cropping;
>
> ☆ Adopting measures for *better use of inputs* (land, labour, capital, water), natural resources (sunshine, rainwater, temperature, humidity), transport and marketing facilities;
>
> ☆ Proper *soil management* by tillage, leveling of land, making of irrigation and drainage channels together with proper checking of soil erosion;

☆ *Nutrient management* by timely application of manures and fertilizers to improve soil fertility and productivity. Efforts to increase soil organic matter by way of supplying the green manure, farmyard manure, organic waste, and biofertilizers;

☆ *Water management* keeping in view the crop, type of soil and environment and conserving the soil moisture and the excess water. Proper scheduling of irrigation at critical stages of crop growth;

☆ Adopting proper *plant protection measures* against weeds, insect-pests, pathogens, and climatic hazards;

☆ Adopting *inter-culture operations*; and

☆ Avoid/reduce harvest losses of crop by harvesting at proper time and using suitable method of harvesting. Proper post-harvest technologies may also be adopted to avoid losses and for making proper use of the crop product.

3. Scope of Agronomy

In view of the over growth of human population and shrinkage of land as a result of acquiring of land for colonization (Urbanization) and establishment of industries, the scope of agronomy has increased with a view to provide food security to every citizen of the country. The agronomy as a dynamic subject has made advancement in knowledge and understanding the soil-plant-environment relationship and development of agro-techniques have been made in the recent past to enhance food production. Still there is a need to develop new and advance practices.

The yield of any field crop depends on different yield attributing characters. This ultimately influences the productivity per unit area of a crop and also the economics of crop production. The yield and its contributing characters indicate the phenotypic value (P) of any particular character which depends on their genotypic values (G) and the environmental effects (E). The *phenotypic value* of any of these characters (P) has thus two main components *viz. genotypic value*(G) and *environmental effects* (E). Thus, P = G + E.

The genotype of the character assigns a value to the character called as genotypic value (G). This genotypic value is modified before final expression of the character by the variation in environmental factors (E). The various environmental factors include soil inherent fertility, different inputs, climatic conditions, and various agronomical/management practices followed regarding the management of field crops in a better way (tillage operations, sowing time, spacing, water management, weed control, fertilizer application, *etc.*). Moreover, the total production of a crop in any state or country can be increased by extending the area under that crop with facilities. This is not possible due to high population pressure on land and consequent colonization and industrialization, particularly in India.

The increase in agriculture production has been restricted by a number of factors *viz.* low genetic potential of crop varieties, low inherent soil fertility due to its depletion by intensive cropping and shortage of organic manures; poor agronomical practices particularly in farmer's field conditions; inadequate control of weeds, insect-pests, and diseases; farmer's poverty to spent money on inputs and hence non-availability of production inputs at farmer's level; no initiative under government economic policies and low price of farm produce; weak research and extension programmes, *etc.* Therefore, evolving and introducing of high yielding improved varieties suited to particular agro-climatic conditions with inbuilt resistance to various stresses and adoption of improved technologies of the package of agronomical practices is the better way of increasing the productivity.

In view of the food security in the country, the agricultural practices have been modified for high productivity, *viz.* availability of chemical as well as biofertilizers, the quantity applied, method and time of their application; availability of herbicides to control weeds, knowing their selectivity, method and time of their application.

Similarly, the water management practices have been developed and requires for further improvement. New technologies have come up to overcome the effect of moisture stress for dry land farming.

Intensive cropping systems (more number of crops in a year on the same piece of land) also needed to develop.

New package of practices are required to be developed to exploit the full genetic potential of new varieties of field crops.

4. Role of Agronomy

The agronomy as an art and science of field crop production and management of soil for higher productivity has significant role in growing of crop plants and their products as under:

    ☆ To provide food for man as well as for animals by cultivating *cereal crops* for their edible starchy grains of large size such as wheat, rice, maize, barley, sorghum, gram,

    ☆ To produce cereals of small grains called *millets* (plants bearing grains like pearls) like pearlmillet (bajra), jawar, shyama, ragi (Finger millet) used as staple food in drier regions,

    ☆ To produce pulse crops which are leguminous crop plants whose seeds are used as food. *Dal* is produced on their splitting and it is rich in protein like, urdbean, mungbean, peas, chickpea, pigeonpea, cowpea, lentil, lathyrus,

    ☆ To provide seeds rich in fatty acids and used to extract vegetable oil for meeting various requirements. These are called *oilseed crops* like mustard, rapeseed, rai, raya, taramira, seasame, linseed, groundnut, sunflower, safflower, soyabean, castor, niger,

☆ To produce sugar and *starch crops* like sugarcane, sugarbeat, potato, sweet potato, tapioca and asparagus,

☆ To produce *beverage crops* whose products are used for mild, agreeable and stimulating liquors meant for drinking like tea, coffee, cocoa,

☆ To produce *spices* and *condiments* whose plants or their products are used to flavor, taste and add zest (lelish – enjoyment of food or savour), and sometimes give colour to the fresh or preserved food like ginger, garlic, onio, turmeric, chillies, coriander,cumin, asofitida, and anise,

☆ To provide fodder for animals by growing of *forage crops* like berseem, cow pea, sorghum, guar, Lucerne, maize, oats, anjan grass, napier (elephant grass), para grass, Guinea grass, Rhodes grass, Sudan grass, Dinanath grass, Teosine, Setaria, stylo, ricebean, fieldbean, velvetbean,

☆ To provide fiber from *fiber crops* like cotton from which fiber is obtained from seed; jute sunnhemp, flax, Roselle or mesta from whose stem the fiber is obtained; and agave, pineapple from which leaves are used to get fiber,

☆ To grow field crops to prepare medicines/drugs like tobacco, mint,

☆ To provide crop plants or their products used for stimulating, numbing drowsing or relishing effects like tobacco, ganja, opium poppy, and anise.

Agronomy is basically the conversion of environmental inputs (solar energy, CO_2, water, soil nutrients) into economic products in the form of human or animal food or industrial raw material. The crop production is one of the agricultural sciences playing a key role in enhancing the total production by improving productivity resulting in food security.

Agronomy provides knowledge about the crops to be cultivated in a particular climate, in different kind of soil, the water management practices, and management practices to be followed to get the higher productivity. Principles and practices are the two important aspects of crop production. The knowledge of the important principles of the crop production is most essential.

5. Agro-Meteorology

The atmosphere surrounding the earth (extending up to a height of about 1600 km.) constitutes a mixture of gases that are odourless, colourless, and tasteless and held to the earth by force of gravity. Thus, the atmosphere is a gaseous envelop of invisible film of air surrounding the earth with no definite upper boundary layer that fades away gradually into inter-planetary space.

5.1 Meteorology

The meteorology is a science to deal with laws and principles applied to atmospheric phenomenon or it can be said as a science of atmosphere with its activities or the meteorology is the science to study weather. Weather is the day to day condition of the air around us and can be described as wet or fine, warm or cold, windy or calm. The measurement of the weather is necessary for farming, sport, transport, the amount of energy used, work, tourism, *etc.*

Elements of Weather

The atmosphere has its influence on climatic changes forming the state of weather at any particular time and place. The various *elements of weather* are:

Solar radiation, air temperature (the amount of heat), atmospheric pressure or air pressure, wind movement (speed and direction), clouds, relative humidity, and precipitation (water – rain, sleet, hail, and snow),and visibility (ability to see).

☆ *Temperature*: This is a measure of hot or colds. It can tell this by looking at the clothes that people wear.

☆ *Precipitation*: Water in the air falls to the ground in one of several forms *viz.* Rain, Snow, Sleet, Hail, also ice, and fog.

☆ *Wind speed*: This tells the velocity of wind. It can get good ideas of this by looking at smoke and trees.

☆ *Wind direction*: This is the direction from which the wind blows- it is measured by a wind vane.

☆ *Cloud*: Clouds come in many shapes, sizes and heights- more about that next week. Cloud cover is the amount of the sky covered by cloud.

☆ *Visibility*: This is the distance that can be seen. It is measured in meters.

Therefore, *meteorology* (study of weather) covers the measurement of temperature, rainfall, air pressure, humidity, sunshine and cloudiness and to make predictions and forecasts about the weather in coming future.

Weather Instruments

Rain Gauge: It is used to gather and measure the amount of liquid precipitation over a set period of time. Most rain gauges generally measure the precipitation in millimeters.

Thermometer: It is a device that measures temperature. It measures the temperature either in Fahrenheit or Celsius. The difference between the daily maximum and the daily minimum temperature is the diurnal range.

Barometer: It is an instrument used to measure atmospheric pressure. Atmospheric pressure is the pressure at any point in the Earth's atmosphere. Although air is very light, because the atmosphere is so thick (many kilometers in altitude above the Earth's surface), air exerts a force or pressure. When air is cooled it sinks towards the ground, the pressure increases and a high pressure is measured. When the air warms up it rises, the pressure is reduced and low pressure is measured. High pressure gives clear sky which means it is hot in the summer and cold in the winter. Low pressure gives us cloudy sky and rain.

5.2 Agro-meteorology

The agro-meteorology is a science to investigate the meteorological, climatologic, and hydrologic conditions that are significant for agriculture. The field of agro-meteorology has its practical utility in protecting or avoiding the adverse climatic risks.

5.3 Agro-climatologic Zones of India

India is a diversified country in all spheres of life than most of the countries of the world with respect to biodiversity of both plant and livestock species, varied agricultural production systems (dry land agriculture as well as irrigated crop cultivation, mixed farming and mixed farming systems), varied animal production system (extensive or landless or zero input, semi-intensive or medium input system, and intensive or high input system of rearing livestock), varied *topography* (hills, plains, rivers, lake, and sea), varied *climatic conditions* (temperate, sub-temperate, semi-arid and arid zone or hot dry climate), various types of soils (sandy, loam, semi-loan, black, red, rocky, and various types of valuable stones, coal mines and various ores or mineral sources), varied seasons for temperature and rainfall variation (summer, rain, autumn, and winter), and cultivation of different crops and raising of different livestock species in different parts of the country.

In view of all these diversities, India has been divided into the following 15 agro-climatologic zones based on the homogeneity in agro-climatic conditions within the zone. These 15 zones can be broadly grouped in to 7 major zones as under:

1. Himalayan Zones – Western and Eastern zones;
2. Gangatic plain zones – Lower, Middle, Upper, and Trans-Gangetic zones;
3. Plateau and hills zones – Eastern, Central, Western, Southern zones;
4. Coast plains and hill regions – East, and West;
5. Gujarat plains and hill region;
6. Western dry region; and
7. Island region.

The characteristics of these zones including the area (districts and states – sub zones) covered under each of these 17 zones, type of soil found in the sub-zone, average rainfall, and major crops of these sub-zones are given here as under –

1. *Western Himalayan zone*: This zone includes J&K, H. P., and Uttarakhand. The soil is of cold region, podsolic soil, mountain meadow soils, hilly brown soils, prone to erosion hazards, and steep slopes in undulating terrain. Average annual rainfall of this region is 1650- 2000 mm. The important crops of this region are wheat, rice, maize, potato.

2. *Eastern Himalayan zone*: This zone consists of Dajeeling hills, Jalpaiguri, and Coochbihar districts of W.B.; and all eastern states (Assam; Sikkim;

Arunachal Pradesh; Manipur; Mizoram; Meghalya; Nagaland; and Tripura). This region is covered with high forest cover. In about 1/3 of cultivated land is under shifting cultivation. Heavy rainfall has caused soil erosion and floods in the lower basins. Average annual rainfall of this region is 1850- 3500 mm. The important crops of this region are rice, jute, maize, and mustard.

3. *Lower Gangetic plains zone*: This zone comprises west-lower gangetic plains region (W. B.). There is alluvial soil in this region. Average annual rainfall of this region is 1300- 1600 mm. The important crops of this region are wheat, rice, jute, mustard.

4. *Middle Gangetic plains zone*: This includes 12 districts of eastern U. P., Bihar, and Jharkhand with high rainfall and its about 39 per cent of gross cropped area is irrigated. Average annual rainfall of this region is 1210-1470 mm. The important crops of this region are wheat, rice, maize.

5. *Upper Gangetic plains zone*: There are 32 districts of U. P. in this zone. There are canals and tube-wells as the irrigation sources with good potential to exploit under-ground water. Average annual rainfall of this region is 720- 908 mm. The important crops of this region are wheat, rice, maize, arhar sugarcane.

6. *Trans-Gangetic plains zone*: This zone comprises Delhi, Punjab and Haryana including Chandigarh and north Rajasthan (Srigananagar district). This area has some peculiarities of highest net sown area, highest irrigated area with high cropping intensity and high ground-water utilization. Average annual rainfall of this region is 360- 900 mm. The important crops of this region are wheat, rice, maize, sugarcane.

7. *Eastern plateau and hills zone*: Southern part of W. B., Odisha, and M. P. are included in this zone. The topography is undulating with a slop of 1 to 10 per cent. The soil is shallow and medium in depth with irrigation by tube-wells and tanks. Average annual rainfall of this region is 1200-1400 mm. The important crops of this region are rice, maize, ragi.

8. *Central plateau and hills zone*: This zone comprises some parts of M.P., Rajasthan. Average annual rainfall of this region is 500- 900 mm. The important crops of this region are rice, jowar, bajra, and soybean.

9. *Western plateau and hills zone*: This zone covers major parts of Maharashtra, parts of M. P. and one district of Rajasthan. The net sown area is about 65 per cent and about 1 per cent forest area. Irrigated area is only 12.5 per cent with canals as main source. Average annual rainfall of this region is 60- 1000 mm. The important crops of this region are wheat, jowar, cotton, sugarcane.

10. *Southern plateau and hills zone* (Semi-arid zone): Some part (35 districts) of Andhra Pradesh, Karnataka, and Tamil Nadu. Dry farming is practiced in about 80 per cent area with a cropping intensity of about 110 per cent.

Average annual rainfall of this region is 600- 1100 mm. The important crops of this region are rice, ragi, and groundnut.

11. *East coast plains and hill regions*: This zone includes east coast of T. N., A. P., and Orissa. The soil of this area is mainly alluvial and coastal sands. Irrigation is by canal and tanks. Average annual rainfall of this region is 800- 1200 mm. The important crops of this region are rice, groundnut.

12. *West coast plains and hill regions*: The zone comprises west coast of T. N., Kerala, Karnataka, Maharashtra and Goa. Average annual rainfall of this region is 2200- 3500 mm. The important crops of this region are rice, ragi.

13. *Gujarat plains and hill region*: There are 19 districts of Gujarat in this zone. It is an arid zone with low rainfall. There is only about 1/3 area which is irrigated by wells and tube-wells. Average annual rainfall of this region is 400- 1400 mm. The important crops of this region are wheat, rice, cotton, and groundnut.

14. *Western dry region*: This zone has 9 districts of Rajasthan and is hot sandy desert, erratic rainfall, high evaporation, and scanty vegetation. The ground water is deep and often brackish (salty). The common features of the region are famine and drought. Average annual rainfall of this region is 350 mm. The important crops of this region are bajra, gram, and mustard.

15. *Island region*: This comprises island territories of Andman and Nicobar, and Lakshadweep. It is typically equatorial with rainfall spread during 8 – 9 months. There are mainly forests in this zone and the land is undulated. Average annual rainfall of this region is 1500- 3000 mm. The important crop of this region is coconut.

Chapter 2
Seeds and Sowing

Seed

☆ Seed may be defined as a fertilized ovule consisting of intact embryo, stored food and seed-coat which is viable and has got the capacity to germinate.

☆ Seed is any material used for planning and propagation whether it is in the form of seed (grain) of food, fodder, fiber or vegetable crop or seedlings, tubers, bulbs, rhizomes, roots, cuttings, grafts or other vegetatively propagated material.

Seeds are the method of pollination and reproduction for flowering plants which include everything from trees to grasses to bushes. Each seed is capable of growing into a new plant, given the right environmental and climatic conditions, and the vast majority of plants have seeds. The choice of a proper seed multiplication model is the key to the further success of a seed program for efficient and maximum multiplication of seeds.

☆ A true seed is defined as a fertilized mature ovule that possesses embryonic plant, stored material, and a protective coat or coats. Seed is the reproductive structure characteristic of all phanerogams. The structure of seeds may be studied in such common types of pea, gram, bean almond or sunflower.

☆ They are all built on the same plan although there may be differences' in the shape or size of the seed the relative proportion of various parts.

First Private Seed Company

Sutton and Sons (existence in 1912 in Kolkata)

Characteristics of the Seed

☆ It must be true to its type.

☆ The seeds must be healthy, pure and free from all inert materials and weed seeds.

☆ The seeds must be viable.

☆ The germination capacity is up to the standard and it has been tested recently.

☆ The seeds must be uniform in its texture, structure and look.

☆ The seeds should be truthfully, labelled and produced under all due cares.

☆ The seed must not be affected by any seed-borne disease.

☆ It should be dry and not mouldy and should contain 12-14 per cent moisture.

Seed Health

Seed health refers that the seed is free from any kind of disease propagules (spore, mycelium, *etc.*), either active or latent.

Classification of Seed

There are four major different classes of seeds in India namely nucleus seed, breeder seed, foundation seed, and certified seed. The different classes of seeds are necessary for the multiplication of quality seed under the vigilant supervision of plant breeder or seed certification agency to distribute seed of notified varieties for sowing purposes to farmers.

Nucleus Seed

☆ The process of development of certified seed for distribution to the farmer of a distant variety is called nucleus seed. This is the cent percent pure seed at genetic and physical levels produced by the plant breeder who evolved the variety without any impurity.

☆ The seed is produced strictly under isolation and maintained by the institute that developed the variety. The seed is made available to different agencies on demand to start seed multiplication chain in the university where breeder seed is produced year after year. This type of seed is not certified by any certification agency.

☆ Vigour and viability of the original variety has to be retained in the nucleus seed. Usually, the pedigree certificate is issued by the concerned breeder after the production of nucleus seed.

Breeder Seed

☆ Breeder seed is the progeny of the nucleus seed which is generally multiplied in a larger area of the field under the supervision of plant breeder and monitored by the breeder seed monitoring committee.

☆ It provides cent percent genetic and physical purity for the production of the foundation seeds. The golden yellow tag is issued for this category by the producing agency. The monitoring committee consists of representatives of state seed certification agencies, national or state seed corporations, ICAR nominee and the concerned breeder.

☆ The breeder stage seed is the first stage seed in the generation system of multiplication. The breeder seed provides the source of first seed, and the subsequent increase in the production of foundation seeds.

☆ The size of the breeder seed tag is 12×6 cm. One tag is generally issued for each and every bag of seeds. The label contains the information like label number, crop, variety, class of seed, lot number, date of test, pure seed percent, inert matter percent, germination percent and producing institution.

☆ Each and every year, the breeder seed price must be decided by the central government uniformly throughout the country. This type of breeder seed is not generally certified by any certification agency.

Foundation Seed

☆ Foundation seed is the progeny of the breeder seed which is handled by the recognized seed producing agencies in public and private sectors under the supervision of Seed Certification Agency in such a way that its quality is maintained according to the prescribed seed standards.

☆ Usually, seed certification agencies issue a white colour tag for foundation seed class. The size of the foundation seed label is 15×7.5 cm. Foundation seed is produced at the State Farm Corporation of India, National Seed Corporation and State seed Corporation under the technical and proper control of qualified plant breeders approved by the Government of India.

☆ Foundation seeds from the interested seed growers. The genetic purity of foundation seed is 99.5 percent. This foundation seeds become the source of all other certified seed classes, either directly or through registered seed producing agencies and hence it is also known as mother seed.

❖ Foundation seed stage I: The foundation seed produced from Breeder seed.

❖ Foundation seed stage II: The foundation seed produced from foundation seed-stage I. It is used for the production of certified category of seed.

Certified Seed

☆ Certified seed is the progeny of foundation seed which is either produced by the registered and certified seed growers under the supervision of Seed Certification Agency by maintaining the quality of certified seed as per the Indian Seed Certification Standards.

✰ Usually, seed certification agencies issue azure blue tag for certified seed class. The size of the certified seed tag is 15*7.5 cm. Genetic purity shows that the seed is of the variety under certification and that there are no admixtures from other varieties or any other crops.

✰ The genetic purity of certified seed is very high with the amount of contamination permitted ranges from 0 – 0.1 percent. A high percentage of germination is necessary from the certified seeds to obtain a good crop stand with the minimum amount of seed.

❖ Certified seed stage I: certified seed produced from Breeder seed I or breeder seed II

❖ Certified seed stage II: certified seed produced from certified seed stage I. It is generally permitted when the reproduction does not exceed three generations excluding breeder seed.

Truthfully Labelled Seeds

✰ The seed is sold based on the result of laboratory established by the producer is called as truthfully labelled seeds. This can be produced and sold by private agencies. In government sectors, many times seed is not able to fulfill seed certification standard for any one parameter.

✰ The colour of truthfully labelled seed tag is an opal green. This seed may be sold by making corrections after getting certification by the authorized seed certification agency.

✰ The price of the truthfully labelled seed is always lower than the certified seed in government sector. Seed rejected due to genetic impurity or presence of the objectionable disease is not labelled as truthful.

Difference between Certified Seed and Truthful Labelled Seed

Certified Seed	Truthful Labelled Seed
Certification is voluntary. Quality guaranteed by certification agency.	Truthful labelling is compulsory for notified kind of varieties. Quality guaranteed by producing agency
Applicable to notified kinds only	Applicable to both notified and released varieties
It should satisfy both minimum field and seed standards	Tested for physical purity and germination
Seed certification officer, seed inspectors can take samples for inspection	Seed inspectors alone can take samples for checking the seed quality.

Source: SeedNet India portal.

Structure of Seed

✰ The various parts of a seed may be easily studied after it has been soaked in water for a day or so varying according to the nature of the seeds. A mature seed contains an embryonic plant (with a radicle and plumule), and is provided with reserve food materials and protective seed coats. A mature pod of pea (*Pisum sativum*) has a number of seeds arranged in two rows.

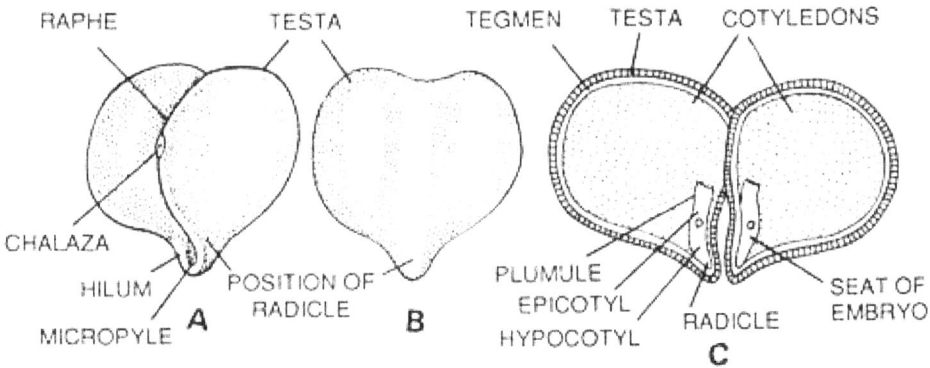

Structure of Gram Seed.
A. Complete seed with various parts B. Seed after removing the testa C. L. S. of seed.

☆ The seeds are attached to the fruit wall by a small stalk, the funiculus. At maturity, on one side of the seed coat a narrow, elongated scar representing the point of attachment of seed to its stalk is distinctly seen, this is the hilum. Close to the hilum situated at one end of it there is a minute pore, micropyle. During seed germination, water is absorbed mainly through this pore, and the radicle comes out through it.

☆ Continuous with the hilum there is sort of ridge in the seed coat, the raphe. The seed is covered by two distinct seed coats; the outer whitish one is the testa, while the other inner thin, hyaline and membranous covering is the tegmen. The seed coats give necessary protection to the embryo which lies within.

☆ The whitish fleshy body, as seen after removing the seed-coats is the embryo. It consists of two fleshy cotyledons and a short axis to which the cotyledons remain attached. The position of the axis lying outside the cotyledons, bent inward and directed towards the micropyle is the radicle and the other portion of the axis lying in between the two cotyledons is the plumule.

☆ The plumule is crowned by some minute young leaves. The radicle gives rise to the root, the plumule to the shoot and the cotyledons store up food material. Since the reserve food material is stored in the massive cotyledons and the seed lacks a special nutritive tissue, the endosperm.

☆ The seeds which lack endosperm at maturity are called non-endospermous or exalbuminous. On the other hand in several other plants such as castor bean (*Ricinus communis*), coconut (*Cocos nucifera*) and cereals, food is stored in the endosperm. Such seeds where endosperm persists and nourishes the seedling during the initial stages are called endospermous or albuminous.

☆ Besides the basic structures (endosperm, embryo and seed-coat) certain special structures may arise during seed development. In castor bean a fleshy whitish tissue, the caruncle, develops at one end of the seed. It is derived from the integument.

☆ The juicy edible part of the litchi fruit (aril) is an outgrowth of the funiculus that develops after fertilization. The cotton fibres are the elongated epidermal cells of the seed-coat. These fibres are single-celled and thin walled. They attain a length of upto 45 mm and have characteristic twists.

General Structure

A seed is generally made up of seed coat and embryo.

Seed Coat

It is the protective covering of the seed derived from one or both integuments of the ovule. Usually seed coat is two layered. The outer thick and hard layer is called testa, while the thin inner membranous layer is called tegmen. The exterior of a seed coat contains prominent structures like hilum, raphe, micropyle and chalaza.

Embryo

It represents the dormant future plant that remains enclosed within the seed coat. Embryo consists of an embryo axis or tigellum and cotyledons. The embryo axis consists of plumule, epicotyl, cotyledonary node, hypocotyle and radicle. Plumule represents the embryonic shoot while radicle represents embryonic root. The embryo axis has a node called cotyledonary node that bears one or two cotyledons (= seed leaves). The part of embryo axis between plumule and cotyledonary node is called epicotyle and a similar region between cotyledonary node and radicle is called hypocotyle. The seed contains reserve food either in cotyledons or in a special tissue called endosperms.

Seed Germination

☆ Germination is the emergence and development of seedlings from the seed-embryo which is able to produce a normal plant under favorable condition.

☆ Agronomically, germination means the capacity of seeds to give rise to normal sprouts within a definite period fixed for each crop under optimum field conditions.

Changes during Germination

☆ Swelling of seed due to imbibition of water by osmosis.

★ Initiation of physiological activities such as respiration and secretion of enzyme.

★ Digestion of stored food by enzymes.

★ Translocation and assimilation of soluble food.

Types of Germination

When seed is placed in soil gets favorable conditions, radical grows vigorously and comes out through micro Pyle and fixes seed in the soil. Then either hypo or epicotyls begins to grow.

★ ***Hypogeal germination***: The cotyledons remain under the soil. *e.g.*: cereals, gram, arhar, lentil.

★ ***Epigeal germination:*** The cotyledons pushed above the soil surface. *e.g.*: mustard, tamarind, sunflower, castor, onion.

Factors Affecting the Germination

Some of the important factors are:

★ External factors such as water, oxygen and suitable temperature.

★ Internal factors such as seed dormancy due to internal conditions and its release.

I. External Factors

★ ***Water:*** A dormant seed is generally dehydrated and contains hardly 6-15 per cent water in its living cells. The active cells, however, require about 75-95 per cent of water for carrying out their metabolism. Therefore, the dormant seeds must absorb external water to become active and show germination. Besides providing the necessary hydration for the vital activities of protoplasm, water softens the seed coats, causes their rupturing, increases permeability of seeds, and converts the insoluble food into soluble form for its translocation to the embryo. Water also brings in the dissolved oxygen for use by the growing embryo.

★ ***Oxygen:*** Oxygen is necessary for respiration which releases the energy needed for growth. Germinating seeds respire very actively and need sufficient oxygen. The germinating seeds obtain this oxygen from the air contained in the soil. It is for this reason that most seeds sown deeper in the soil or in water-logged soils (*i.e.* oxygen deficient) often fail to germinate due to insufficient oxygen. Ploughing and hoeing aerate the soil and facilitate good germination.

★ ***Suitable Temperature:*** Moderate warmth is necessary for the vital activities of protoplasm, and, therefore, for seed germination. Though germination can take place over a wide range of temperature (5-40°C), the optimum for most of the crop plants is around 25-30°C. The germination in most cases stops at 0°C and 45°C.

II. Internal Factors

☆ *Seed Dormancy Due to Internal Conditions and Its Release:* In some plants the embryo is not fully mature at the time of seed shedding. Such seeds do not germinate till the embryo attains maturity. The freshly shed seed in certain plants may not have sufficient amounts of growth hormones required for the growth of embryo. These seeds require some interval of time during which the hormones get synthesized.

The seeds of almost all the plants remain viable or living for a specific period of time. This viability period ranges from a few weeks to many years. Seeds of Lotus have the maximum viability period of 1000 years. Seeds germinate before the ending of their viability periods. In many plants, the freshly shed seeds become dormant due to various reasons like the presence of hard, tough and impermeable seed coats, presence of growth inhibitors and the deficiency of sufficient amounts of food, minerals and enzymes, *etc.*

Methods for Testing Germination

☆ *Petridish Method:* Two blotters or filter papers are placed on the bottom of the petridish and they are soaked with water. A convenient number of seeds, ranging from 10-20, are placed on the surface of water-soaked blotters in the petridish. The kind of seed, date and time of seed soaking are written on the glass-cover of petridish with the help of a glass marking pencil. Usually, the germination percentage is calculated in two counts and reported on the basis of the results of germination of about 100 to 200 seeds. This method is suitable for small seeds *viz.* tobacco, tomato, radish, knol-khol, cabbage, cauliflower, mustard, lettuce, brinjal, chillies, *etc.*

☆ *Rolled Towel Method:* Two wet towels are placed on a smooth table top. The appropriate number of seeds are placed on the upper surface of the towels and are covered by two wet towels. A fold is made of the bottom of the towel to prevent the seeds from falling out. The towels are then rolled from right to left. The full informations regarding the test *i.e.* the kind of seed, lot number, date and time of seed soaking are noted on the roll with the help of an indelible pencil. This method is suitable for comparatively large sized seeds *viz.* maize, wheat, pea and gram.

☆ *Folder Paper Towel Method:* Two wet paper towels (big bamboo sheets), specially made for germination test, are placed on the working table-top. The surface of paper-towel is marked into two halves. The right half of the towel surface is planted with seeds and the left side half is folded to cover the right half and informations regarding the seed sample, date and time of seed soaking are written on the fold with an indelible pencil. Water is frequently sprinkled on these towels and observations on germinating seedlings and their numbers are taken periodically. This method of germination test is convenient for relatively large sized seeds.

☆ ***Sand Method:*** Seeds are planted in uniform layer of moist sand and then covered to depth of 1-2 cm with sand. Substrates for germination are moistened with 0.2 per cent KNO_3 solution.

☆ ***Rags or Gunny Sacs Method:*** The seeds are simply wrapped in a moistened rag or gunny sac, which is then rolled and tied loosely in the form of bundle. This is now kept at a proper temperature on a rack or convenient place for germination. The rolls are opened after a fixed period of time and the number of germinated seeds is counted.

☆ ***Mechanical Method:*** In this method, apparatus like Germinators are used. Cabinets of the incubator type with thermostatically controlled temperature may be used for the purpose. In these cabinets the seeds may be placed evenly on moist filter paper in petri dishes or between filter papers kept moist by folds of moistened flannel or large seeds may be sown in dishes containing sand or fine soil. Adequate water is applied when needed by the help of a wash-bottle or sprayer.

Seed Purity

The purity of seed denotes the real percentage of desirable seed from a lot of seed with several impurities *i.e.* seeds of other varieties, other crop seed, defective seeds, weeds seeds, inert matter, sand straw, stones, soil and iron particles, pebbles *etc.* Dockage is the impurity percentage of seed.

Real Value of Seed

The real value of seed represents its seed quality in terms of purity and germination. It can be evaluated by multiplying the purity percentage and germination percentage of a seed sample and dividing the product by 100. Genetic purity is tested by field plot test.

Viability Test of Seed

☆ Viability of seeds represents the capacity of the seed to germinate.

☆ Viability is defined as the capacity of the seed to remain capable of germination for some specific period of the time.

☆ Seed vigour is the ability of seed to emerge in varying environments of fields.

Methods for Viability Test of Seeds

Electrical Conductance Method

☆ The seeds are soaked in distilled water and the electrical conductance of the bathing solution is tested, the increase in conductance is roughly proportional to the percentage of dead tissues.

☆ The increase in the conductivity is due to leaching of metabolites from dead seeds which become previous owing to increasing permeability.

Potassium Permanganate Method

☆ The seeds are soaked in a weak solution of potassium permanganate and as the proportion of dead seed increases the discolouration of the solution also increase.

☆ The dead cells become freely permeable to their contained solutes which leach out easily into bathing solution, whereas leaching from living cells is very less.

☆ Thus, extent of discolouration indicates the proportion of dead seeds into a seed-lot.

Indigo-carmine Method

☆ The seeds are soaked in the solution of any aniline dyes such as indigo-carmine for few hours, it is observed that the dead seeds or their cells are stained.

☆ Thus, the portion of dead and viable seeds can be determined by counting the number of stained and unstained seeds respectively.

Embryo Culture Method

☆ In this method, the embryo is removed carefully from its cotyledons or endosperm and then it is placed naked on granulated peat mass or on sterilized nutrient agar medium (White's nutrient medium).

☆ The method takes about 7 to 10 days to give the viability percentage of the seeds.

☆ The viability is judged by counting the number of germinated embryos because the viable seeds will germinate and non-viable will fail to do so.

Tetrazolium Chloride Test

☆ The tetrazolium chloride test is also known as "Biochemical test".

☆ In this method the seeds are soaked in 0.5 to 2 per cent solution of tetrazolium chloride (2, 3, S-triphenyl tetrazolium chloride).

☆ The viable or living seeds take bright red colouration which becomes more intense in the embryo while the dead seeds remain in their original colour.

☆ The method can be used as a guide in blending seeds and their lots under an urgent and immediate demand for seed.

☆ First Seed Testing Laboratory was established at IARI in 1961.

Pure Live Seed

Isolation Distance

Isolation distance is a separation distance of pure and healthy plants from that of contaminated plants to avoid the contamination or cross pollination.

Sl.No.	Crop	Distance for Foundation Seed (m)	Distance for Certified Seed (m)
1.	**Self pollinated crops**		
	Rice, Wheat, Groundnut	3	3
	Moong, Cowpea, Tomato	50	50
2.	**Cross pollinated crops**		
	Maize, Mustard	400	200
	Pearlmillet	1000	25
	Sunflower	1000	500
	Castor	300	150
	Sunhemp	200	100
	Lucerne	400	100
	Onion	1000	400
3.	**Often cross-pollinated crop**		
	Pigeonpea	100	50
	Cotton	50	30
	Sorghum	200	100
	Chilli, Okra	400	200
	Brinjal	200	100

Test weight - weight of 1000 seeds

Seed Index - weight of 100 seeds (normally for bold seeded crops)

Seed Dormancy

Introduction

☆ Dormancy is the state of rest period of a seed in which it does not germinate or it is a state of inhibited growth of seed or other plant organs as a result of internal causes but the seed remains viable for

Seed dormancy can be defined as the state or a condition in which seeds are prevented from germinating even under the favourable environmental conditions for germination including, temperature, water, light, gas, seed coats, and other mechanical restrictions.

a period upto few years. Rice varieties Masoorie, I.R.-50 have no seed dormancy.

☆ The main reason behind these conditions is that they require a period of rest before being capable of germination. These conditions may vary from days to months and even years. These conditions are the combination of light, water, heat, gases, seed coats and hormone structures

Importance of Seed Dormancy

☆ It follows the storage of seeds for later use by animals and man.

☆ It helps in the dispersal of the seeds through the unfavourable environment.

☆ Dormancy induced by the inhibitors present in the seed coats is highly useful to desert plants.

☆ Allows the seeds to continue to be in suspended animation without any harm during cold or high summer temperature and even under drought conditions.

☆ Dormancy helps seeds to remain alive in the soil for several years and provides a continuous source of new plants, even when all the mature plants of the area have died down due to natural disasters.

Causes of Dormancy

There are certain major causes for the seed dormancy. Listed below are the few reasons for the seed dormancy.

☆ Seed coats being impermeable to water

☆ Hard seed coat

☆ Seed coats being impermeable to oxygen

☆ Rudimentary embryo of seeds

☆ Dormant embryo

☆ Synthesis and accumulation of germination inhibitors in the seeds

☆ Period after ripening

☆ Germination inhibitors

☆ Immaturity of the seed embryo

☆ Impermeability of seed coat to water

☆ Impermeability of seed coat to oxygen

☆ Mechanically resistant seed coat

☆ Presence of high concentrate solutes

Kinds of Dormancy in Seeds

☆ **Primary dormancy:** The seeds which are capable of germination just after ripening even by providing all the favorable conditions are said to have primary dormancy. *e.g.*: Potato.

☆ **Secondary dormancy:** Some seeds are capable of germination under favourable conditions just after ripening but when these seeds are stored under unfavourable conditions even for few days, they become incapable of germination.

☆ **Special type of dormancy:** Sometimes seeds germinate but the growth of the sprouts is found to be restricted because of a very poor development of roots and coleoptiles.

Methods to Break the Seed Dormancy

I. Scarification

The dormancy due to hard seed coat or impermeable seed coats can be broken by scarification of seed coats. It should be done in such a way that the embryo is not injured.

☆ *Chilling (Pre-chilling):* The seeds are placed in contact with the moist substratum at a temperature of 5 to 10°C for 7 days for germination. *e.g.* Cabbage, Cauliflower, and Sunflower.

☆ *Pre-dying:* Seeds should be dried at a temperature not exceeding 40°C with free circulation for a period of 7 days before they are placed for germination. *e.g.* Maize, Lettuce.

☆ *Pre-washing:* In some seeds, germination is affected by naturally occurring substances which act as inhibitors which can be removed by soaking and washing the seeds in the water before placing for germination. *e.g.* Sugar beet.

☆ *Pre-soaking:* Some seeds fail to germinate due to hard seed coat. Such seeds should be soaked in warm water for some period so as to enhance the process of imbibitions. *e.g.* Chilli, Subabul.

☆ *Rubbing or puncturing seed coat:* Some seeds are subjected to mechanical scarification either by rubbing them against rough surface or puncturing the seed coat with pointed needle. *e.g.* Coriander, Castor.

☆ *Application of pressure to seeds:* Germination of *Medicago sativa* is found to be increased when a hydraulic pressure of 2000 atmosphere at 18°C is applied. It may be due to increase in permeability of seed coat to water and O_2.

II. Stratification

In some seeds after ripening, low temperature and moisture conditions require in artificial stratification. Seed layer altered with layers of moist sand or appropriate material to store at low temperature. *e.g.* Mustard and Groundnut.

III. Exposure of Seeds to Light

It also helps to break the dormancy and increase the germination.

IV. **Chemical Treatments**

    ☆ *Potassium nitrate treatment (KNO₃)*: The material used for placing the seeds for germination *i.e.* substratum, may be moistened with 2 per cent solution of KNO_3 (2g KNO_3 + 100ml of water). *e.g.* rice, tomato, chillies.

    ☆ *Gibberellic acid treatment*: The substratum used for germination may be moistened with 500 ppm solution of GA *i.e.* 500 mg in 1000ml water. *e.g.* Wheat, Oat.

    ☆ *Thiourea treatment*: Potato tubers are dipped in thiourea solution (1 per cent) for one hour when fresh harvested produce is to be used as seed material.

Seed Treatment

Seed treatment is a process of application of chemicals or protectants (with fungicides or bactericides or nematicides) to seeds that prevent the carriage of insects or diseases causing pathogens in/on the seeds. Seed treatment also enables the seed to overcome seedling infection by soil-borne fungi.

Objects of Seed Treatment

    ☆ **To control disease:** by treated the seeds with fungicides or organomercurial compounds like Thiram, captain, carbendazim, agrosan, cereson, *etc. e.g.* to control paddy blast, seed is to be treated with agrosan @ 3 g per kg (3g/kg) of seed).

    ☆ **To have convenience in sowing** due to special characteristics of the seed like fuzz of cotton seeds, coriander seeds, small seeds of chilli, ragi, bajara, *etc. E.g.* coriander seed is to be splitted by rubbing it against hard surface. Seed of chilli, Sesamum, bajara are mixed with fine sand or soil.

    ☆ **To have quick germination** with broken thick seed coat by mixing them with coarse gritty sand and trampling or pounding it lightly in a morter with a wooden pestle for breaking the thick seed coat or soaked the seeds in water for a specified time. *e.g.*: cotton seed or paddy seed is soaked in water before actual sowing.

    ☆ **To increase nitrogen fixation in legumes:** Legume seeds are inoculated with a particular *Rhizobium* culture. This is mixed with jaggery solution and applied to seed and dried in shade. It increases nodulation and thereby N fixation.

    ☆ **To protect the seed against insect pests**:

    ☆ **To induce earliness (Vernalization treatment).**

    ☆ **To break dormancy** by treating seed with chemicals. *e.g.* Thiourea 1 per cent treatment to potato tubers.

Seed Treatment in Important Crops

1. **Sorghum:** Thiram or 300 mesh sulphur: Seed is coated in seed dressing drum or earthen pot @ 3.4 g/kg seed against smut disease.

2. **Bajara:** Brine solution treatment is given @ 20 per cent against eat got and to discard light and diseased seed.

3. **Paddy:** Seed is treated with brine solution @ 3 per cent against blast of paddy and to discard unfilled seed.

4. **Wheat and Oilseed crops:** Seed is coated with Thiram or Bavistin @ 5 g/kg seed against seed borne diseases.

5. **Cotton:** Seed is treated with organomercurial compound like Ceresan, Agrosan @ 3 g or Thirum @ 5g against seed borne disease like anthracnose.

6. **Small seeded crops** like Sesamum, bajara, tobacco, *etc.*: Seed is mixed with fine sand or soil for even sowing of seed in the field.

7. **Potato**: (*a*) Seed is dipped in 1 per cent Thiourea solution for breaking the seed dormancy. (*b*) Seed is dipped in streptomycin solution @ 200 g in 100 lit. Water for 1 hour against Ring rots disease.

8. **Legume crops** like Mung, Urd, Soybean *etc.* (*a*) Seed is treated with Thiram @ 3 g/kg seed against seed borne disease. (*b*) Seed is treated with *Rhizobium* culture @ 250g/10kg seed for 'N' fixation and better nodulation.

9. **Sugarcane:** (*a*) Hot water treatment (50ºC) or hot air treatment (54ºC) is given to sets for 2 hrs. Against grassy shoot and other diseases. (*b*) Sets are treated with OMC 6 per cent @ 500g in 100 lit. Water by dipping for 5 min. against smut and increase germination or Bavistin @ 200g in 100 lit. for 5 min.

Seed Certification

"Seed certification is a legally sanctioned system for quality control of seed multiplication and production" and which consists the control measures are:

1. **Field inspection:** At the time of growing a crop for seed production purpose. The data should be obtained on trueness to varietals purity, isolation of seed crop to prevent crops- pollination, mechanical admixtures and diseases dissemination, objectionable weeds and admixtures.

2. **Supervision on agricultural operations** *i.e.* intercultural operations, harvesting, storage, transport and processing *etc.* for identity and quality of lots.

3. **Sample inspection:** For quality and to maintain genetic purity, a lab test of representative samples drawn by the S.C.A. for determining, percentage of germination moisture content, weed seed content, admixture and purity.

4. **Bulk inspection:** For checking homogeneity of the bulk as compared with the sample inspected.

5. **Control Plot Testing:** Samples drawn from the source seed and the final seed produced can be grown in the field along with standard samples of the variety.

The purpose of seed certification is to maintain and make available high quality seed and propagating materials of notified plant varieties.

Phases of Seed Certification

   I. Verification of seed source.

   II. Inspection of seed crop in the field.

   III. Supervision at postharvest stages including processing and packing.

   IV. Seed sampling and analysis.

   V. Grant of certificate, certification tag, tables and sealing.

Indian Seed Act, 1966

General Introduction

☆ To ensure the availability of quality seeds, Government of India has enacted Seeds Act, 1966

☆ To regulate the quality of certain notified kind/varieties of seeds for sale

☆ Seeds producer can operate effectively and make good quality seed available to cultivators

☆ The Seed Act was passed by the Indian Parliament on 29th December 1966 and it came into force from 2nd October, 1969

Seed Legislation

☆ Sanctioning legislation: authorizes formation of Advisory bodies, Seed Certification Agencies, Seed Testing laboratories, FS and CS programme, Recognition of Seed certification Agencies of Foreign countries, Appellate authorities *etc.*

☆ Regulatory legislation: controls the quality of seeds sold in the market including suitable agencies for regulating the seed quality.

Short Title, Extent and Commencement

☆ THE SEEDS ACT, 1966 (ACT NO. 54 OF 1966)

☆ Enacted by Parliament in the 17th year of the Republic of India

☆ This act extent to the whole of India

Definitions

"Agriculture" includes horticulture;

"Central Seed Laboratory" means the Central Seed Laboratory established or declared as such under sub-section (I) of section 4;

"Certification agency" means the certification agency established

`Committee" means the Central Seed Committee constituted under sub-section (1) of Section 3;

'Container" means a box, bottle, casket, tin, barrel, case, receptacle, sack, bag, wrapper or other thing in which any article or thing is placed or packed;

'Export" means taking out of India to a place outside India; "Import" means bringing into India from a place outside India;

'Kind" means one or more related species or sub-species of crop plants each individually or collectively known by one common name such as cabbage, maize, paddy and wheat;

Central Seed Committee

CHAIRMAN

8 representatives from Central Govt

2 representatives for seed growers

1 representative from each State Govt

☆ The members of the Committee shall be entitled to hold office for 2 years

☆ Make bye-laws fixing the quorum and regulating its own procedure

☆ Committee may appoint one or more sub-committees

☆ Central Government shall appoint a person to be the secretary of the Committee

Central Seed Laboratory and State Seed Laboratory

☆ Declare any seed laboratory as the Central Seed Laboratory to carry out the functions entrusted as per this act

☆ State Seed Laboratory where analysis of seeds of any notified kind or variety shall be carried out by Seed Analysts

☆ The Seed Testing Laboratory at the IARI, New Delhi, has been notified as the Central Seed Testing Laboratory during 1960.

❖ Initiate testing programme in collaboration with the State Seed Laboratories in India designed to promote uniformity in test results.

❖ Collect data continuously on the quality of seeds found in the market and make this data available to the Committee

❖ Act as referee laboratory in testing seed samples for achieving uniformity in seed testing.

❖ Testing of disputed sample from different state Seed testing laboratory and private seed testing organization.

☆ Central Seed Testing Referral Laboratory was established under NSRTC, Ministry of Agriculture and Farmers Welfare at Varanasi as a separate National Seed Quality Control Laboratory, which serves as a (CSTL) on 1st April, 2007.

☆ **Power to notify kinds or varieties of seeds:** To regulate the quality of seed of any kind or variety to be sold for agriculture purposes, it is to be a notified kind in the Official Gazette.

☆ **Power to specify minimum limits of germination and purity,** *etc.*: The minimum limits of germination and purity with respect to any seed of any notified kind or variety are recommended.

☆ **Regulation of sale of seeds of notified kinds or varieties:** Seed should be notified kind which confirms to the minimum limits of germination and purity sealed in a container with the mark or label containing the correct particulars.

☆ **Certification agency:** Establish a certification agency for the State to carry out the functions.

☆ **Grant of certificate by certification agency:** Any person selling, bartering or supplying any notified kind or variety seeds, if he desires to have such seed certified can apply to the certification agency for the grant of a certificate.

☆ **Revocation of certificate:** If the holder of the certificate without reasonable cause, failed to comply with the conditions or has contravened any of the provisions of this Act, can revoke the certificate.

☆ **Appeal:** Any person aggrieved by a decision of a certification agency may, within 30 days from the date on which the decision is communicated to him, prefer an appeal to the appellate authority.

Seed Analyst

☆ The State Government may, appoint such persons having the prescribed qualifications, to be Seed Analysts and define the areas within which they shall exercise jurisdiction.

☆ On receipt of a sample for analysis, Seed Analyst shall first ascertain the mark and the seal or fastening.

☆ The Seed Analyst shall analyze the samples according to the provisions of the Act and deliver the result report of the analysis.

Seed Inspector

Every Seed Inspector shall be deemed to be a public servant under section 21 of the Indian Penal Code (45 of 1860) and shall be officially subordinate to such authority as the State Government may specify.

Powers of Seed Inspector

- ✫ Inspection of all places used for growing, storage or sale of any seed of any notified kind or variety by certification agency
- ✫ Procure and send samples of any seeds for analysis, if necessary, which may be suspected for being produced, stocked or exhibited for sale in contravention of the Act
- ✫ Investigate any complaint, which may be made to him in writing in respect of any contravention of the provisions of the Act
- ✫ Maintain a record of all inspections made and action taken by him in the performance of his duties including the taking of samples and the seizure of stocks
- ✫ Submit copies of record to the Director of Agriculture or the certification agency (f) Detain imported containers which are suspected to contain seeds, import of which is prohibited except and in accordance with the provisions of the Act (g) Provisions of Criminal Procedure 1898 (5 of 1898) code: search or seizure made under the authority of a warrant issued under section 98.

Procedure to be followed by Seed Inspectors

- ✫ Give notice in writing, then and there, of such intention to the person from whom he intends to take sample
- ✫ When samples of any seed of any notified kind or variety are taken:
  - ❖ Deliver one sample to the person from whom it has been taken;
  - ❖ Send another sample for analysis to the Seed Analyst for the area within which such sample has been taken;
  - ❖ Retain the remaining sample for production in case any legal proceedings are taken or for analysis by the Central Seed Laboratory.
- ✫ if he seizes the stock of the seed, he shall inform a magistrate and take his orders as to the custody.

Report of Seed Analyst

The report sent by the Central Seed Laboratory.

Restriction on Export and Import of Seeds of Notified Kinds or Varieties

- ✫ It conforms to the minimum limits of germination and purity
- ✫ Its container bears mark or label with the correct particulars

Recognition of Seed Certification Agencies of Foreign Countries

The Central Govt. on the recommendation of the Committee recognize any seed certification agency established in any foreign country.

Penalty

☆ For the first offence with fine which may extend to 500 rupees

☆ Imprisonment for a term which may extend to 6 months, or with fine which may extend to 1000 rupees, or both.

☆ **Forfeiture of property:** The seed in respect of which the contravention has been committed may be forfeited to the Government.

☆ **Offences by companies:** Where an offence under this Act has been committed by a company and it is proved with the consent to any neglect on the part of, any director, manager, secretary or other officer of the company, shall be deemed to be guilty of that offence and punished accordingly.

☆ **Protection of action taken in good faith:** No prosecution or other legal proceeding shall lie against the Government or any officer for anything done in good faith or intended to be done under this Act.

☆ **Power to give directions:** The Central Government may give such directions and execute duties to any State Government.

☆ **Exemption:** Any seed of any notified kind or variety grown by a person and sold by him in his own premises direct to another person for the purpose of sowing.

Power to make Rules

☆ Functions of the Central Seed Laboratory;

☆ Functions of a certification agency;

☆ Manner of labelling the container of seed of any notified kind or variety

☆ Form of application for the grant of a certificate under section 9

☆ Form and manner in which and the fee on payment of which an appeal may be preferred

☆ Procedure to be followed by the appellate authority in disposing of the appeal

☆ Qualifications and duties of Seed Analysts and Seed Inspectors

  ❖ The manner in which samples may be taken by the Seed Inspector,

  ❖ The procedure for sending samples to the Seed Analyst or the Central Seed Laboratory

  ❖ Analyzing of seed samples

❖ Form of report of the analysis results

❖ Records to be maintained by a person

The Central Seed Certification Board

Under section 8a of the seed act 1966, which was introduced into the act by THE SEEDS (AMENDMENT) ACT, 1972 (No. 55 of 1972) provides for the establishment of a central seed certification board.

Co-ordinates the Functioning of the Agencies and Certification

☆ a Chairman to be nominated by the Central Government

☆ 4 members, to be nominated by the Central Government employed by the State Governments as Directors of Agriculture

☆ 3 members, to be nominated by the Central Government employed by the Agricultural Universities as Directors of Research

☆ 13 persons, to be nominated by the Central Government of which not less than 4 persons shall be representatives of seed producers or tradesmen.

Sowing of Seed

☆ Sowing is a process of planting seeds into the soil. During this agricultural process, proper precautions should be taken, including the appropriate depth, proper distance maintained, and soil should be clean, healthy and free from disease

Sowing is an art of placing seeds in the soil at particular depth for good germination of the seeds. Sowing plays a major role in Agriculture. Sowing is step followed next to land levelling. Perfect sowing is placing the seed at a specific depth with the correct amount of seed per unit are with good spacing between plant to plant and row to row.

and other pathogens including fungus. All these precautions are essential for **seed germination** – the process of seeds developing into new plants.

☆ Sowing plays an important role in farming. Once, after the soil is loosened and ploughed, the good, disease-free and pure quality of seeds are selected and sown into the soil. After selecting seeds of good quality, they are sown on the prepared land. The seeds, which give high yields are usually selected and are sown by the following methods.

Soil Preparation for Sowing the Seeds

Soil preparation is the first and foremost step to be followed before sowing seeds. There are various methods of land preparation in crop production. The process of soil preparation includes three important steps. They are:

Ploughing

- ☆ This is the first step in soil preparation before sowing. Ploughing is the process of loosening and turning of the soil upside down in order to make sure the availability of nutrients uptake by the plants. Ploughing helps in easy penetration and germination of the seedlings after sowing.

- ☆ It also helps in proper aeration to the plants. Loosening the soil helps in the growth of beneficial microbes and earthworms which are highly beneficial to the crops. In addition, Ploughing helps in weed control by removal of the volunteer hosts or weeds in the crop. Furthermore, ploughing makes the soil nutrient rich.

- ☆ Ploughing is the mandatory agriculture practice followed before sowing almost in all the Agricultural crops. However, the number of ploughings may vary depending on the crop which we select.

- ☆ Ploughing may be done using various agricultural equipment like Plough, hoe, cultivator, *etc.* Ploughing may also be followed by harrowing in hard soil crops with equipment known as a harrow.

Leveling

- ☆ It is step followed next to ploughing. Leveling is nothing but the even spread of the soil after ploughing. Leveling of the land varies with crops.

- ☆ Leveling includes making of ridges, furrows and other designs that suit for a specific crop.

- ☆ Leveling helps in easy distribution of irrigation water in the crops that help in even uptake of water by all the plants in the crop. The equipment used for leveling is known a leveler.

Manuring

- ☆ It is practice followed before sowing. It is the addition of organic amendments like cow dung, earthworms, neem extracts, synthetic fertilizers to soil to enrich the nutrient content if the soil.

- ☆ Before manuring the soil, we should compulsorily go for soil testing in order to ensure the availability of nutrients and minerals at appropriate quantity in the soil.

- ☆ Based on the information available in the soil test report we must go for manuring the soil with natural amendments.

Methods of Sowing

Seeds may be sown directly or they can be transplanted. For transplanting method, the seeds are sown in nursery and then the nursery is transplanted to field. The different methods of sowing are given below:

Broadcasting

Broadcasting is the most common and oldest methods of seed sowing. The seeds are randomly thrown on the prepared land or seedbed with hand physically. However, it can also be done mechanically. Even though, manual broadcasting is beneficial for some crops which are also economical too. While broadcasting the farmer must ensure that the person should be enough skilled to broadcast the seeds. When we go for manual broadcasting, it consumes a high quantity of seed or the seed rate is high in the broadcasting method of seed sowing. However, the machinery used in broadcasting is good and efficient as it scatters the seeds at controlled rates on the field.

Advantages of Manual Broadcasting

Manual method is cheap and it takes less time than other methods. This method is suitable only for small seeded and crops where plant to plant distance is small or does not matter.

Disadvantages of Broadcasting

In this method the Seed distribution is uneven,some seeds are not be covered by soil. The germination of seed is Non-uniform and Crop stand is affected by uneven distribution.

Dibbling

Dibbling is the process in which we place seeds in the holes or pits at equal predetermined distances and depths. This procedure is done by dibble, planter or manually. It involves the placing of the seeds in the holes manually or with specific instruments known as dibblers. Those dibblers can be operated manually by skilled labor or skilled farmers. The dibbler is the conical instrument in order to make proper holes while sowing of the seeds in the field. This method of sowing is mostly used for sowing of the Vegetable Crops.

Advantages

In this method Germination is rapid and uniform, less seeds are required and seedling vigor is good.

Disadvantages

This method is time consuming,costly and More labor is required.

Drilling

In this process the seed is dropped into holes, the seeds are then covered and compacted by soil. Drilling is done with the help of seed drill or seed-cum-fertilizer drill. Seeds are drilled continuously in a row or drilling can be done at distance which is set and rows can be made accordingly.

Advantages

In this method the quantity of seed required is less and at the time of drilling manures, fertilizers and amendments can applied with seeds.

Disadvantages

This method required more labour and it is more time consuming so cost is also very high.

Sowing Behind Country Plough

In this process, the seeds are placed into the furrows ploughed in the field either continuously or at specific distance manually by a man working behind plough. The depth of sowing depends on the depth of plough.

Planting

Planting is the placing of seeds or propagules firmly in the soil for germination and growth.

Transplanting

Transplanting is the method in which planting of seedlings in main field after pulling out from the nursery.

Precautions

There are a few necessary precautions, which need to be followed while sowing the seeds.

- ☆ The seeds should be disease-free.
- ☆ Seeds must be planted at correct distances from each other.
- ☆ Seeds should be sown such that all the crops should get an equal amount of light, nutrients, and water.
- ☆ Seeds should be sown at correct depths. They should neither be placed at the top of the soil so that it is blown away by wind and animals, nor should it be sown too deep into the soil such that it does not germinate.

Chapter 3

Tillage

Tillage is the physical manipulation of soil with tools and implements to result in good tilth for better germination of seed and subsequent growth of crops (Ready *et al.,* 1993). Tillage is the tilling of land for bringing about conditions that are favourable for the cultivation of crops. The word tillage is derived from the Angle-Saxon word *'tilion'* and *'teolian'* meaning to plough and prepare soil for sowing the seed, transplanting the seedlings and raising the crops. Tilth implies to the physical condition of soil in its relation to plant growth. Tilth is brought out by tillage. Tillage is the primary function of cultivation and it is laborious and expensive cultural practice. Tillage helps to replace natural vegetation with useful crops and is necessary to provide a favourable edaphic environment for establishment, growth and yield of crop plants. Tillage helps to improve the physical condition of soil, control of weeds, insect pests and diseases and also bring the nutrient available to plant. The cultivation is not possible without tillage operation. The crops production depends on good tillage operations.

The word of manures is originated from the French word "Manoeuvrer" which refers to 'Work with Soil', that is why the word tillage and manure where synonyms as is clear by the statement of Jethro Jull (1700 B.C.) "Tillage is manure", there must be sufficient moisture in the soil for good tillage. Tillage is quite impossible in fully dry soil. On the other hand, tillage in wetland results in clogging of soil. Tillage in wetland having sufficient water brings puddling condition of soil which is favourable for cultivation of transplanted paddy, onion *etc.*

Objectives of Tillage

There are several objectives of tillage as follows :

1. The main objectives of tillage is the preparation of land that becomes suitable for crop cultivation. The soil becomes hard and compact after

the harvest of crop and this soil is quite unsuitable for crop cultivation. The soil should be loose, and friable with sufficient air and water for germination of seed.

2. Tillage helps to control weeds by uprooting and smothering the young seedling or weeds. Tillage may induce germination of Weed seeds which can be destroyed by subsequent tillage. The grown up weeds may also be controlled by tillage.

3. Tillage improves the soil structure. The soil become loose, friable and granular if it is ploughed properly at optimum moisture level.

4. Tillage helps to incorporate weeds, crop residues, green manures and other organic manures and, fertilisers, soil amendment and other agro-chemicals applied for the control of weeds, insect pests *etc.*

5. Tillage helps to improve the capacity of soil to receive rain or irrigation water and to retain the moisture for crop growth.

6. Tillage improves the soil aeration which is beneficial for respiration of plant root and soil microorganism and their multiplication.

7. Tillage exposes the lower soil to weather and places the surface soil underneath. As a result of which the soil inhabiting organism comes out and destroyed by heat and predatory animals and birds.

8. Tillage improves the soil permeability and help to conserve soil moisture as a result of soil mulch. It also check the loss of water through percolation in lowland paddy field by making an impervious soil layer.

9. Tillage provides condition suitable for growth, nutrition and development of beneficial soil organism.

10. Tillage remove the hard pan that increases the soil depth for water absorption.

11. Tillage helps to modify slightly the thermo capacity of soil.

12. Proper tillage results in soil and moisture conservation through higher infiltration, reduced run-off and increase depth of soil for moisture storage.

13. Organic matter decomposition is hastened resulting in higher nutrient availability if the soil is ploughed properly at optimum moisture level.

14. Tillage makes the soil level and flat that are suitable for uniform movement of irrigation and drainage water.

15. Tillage brings the soil in such a condition that are conductive for root penetration as the root fails to penetrate in hard soil.

16. The most important objective of tillage is to prepare good seed bed. Good seed bed is necessary for early seed germination and initial good stand of the crop.

17. The stubbles of previous crop which harbour insect pests are removed following tillage resulting in reduced pest attack on the succeeding crop.

Types of Tillage

(A) Preparatory Tillage

Tillage operations that are carried out to prepare the field for raising crops from the harvest of a crop to the sowing of the next crop are known as preparatory tillage. It is divided into primary and secondary tillage operations.

(i) Primary Tillage or Ploughing

The tillage operation that is done after the harvest of crop to bring the land under cultivation is known as primary tillage. Ploughing is opening of the compact soil with the help of different ploughs. Primary tillage is done mainly to open the hard soil and to separate the top soil from lower layers and to uprooting of weeds also. This operations consist of ploughing, discing, harrowing and levelling of the field. Country plough, Disc plough, Mould board plough. Bose plough, Tractors, Power tiller *etc.* are used for primary tillage.

Depth of Ploughing

Depth of ploughing may ranges from 10 to 30 cm depending on the root system of the crop that is to be cultivated. Deep ploughing is required for the tap rooted crops while the shallow ploughing required for shallow and fibrous rooted crops.

Time of Ploughing

Soil moisture determine the optimum time of ploughing. When the soil, particularly heavy soil is dry, it is difficult to open the soil and great clods are turned up, if ploughing is done by any means which is difficult to work into a good seed bed. On the other hand, the soils, when they are too wet are ploughed, get puddled and becomes impervious to air and water and also structureless. This soil becomes hard on drying and unworkable untill remoistened.

Number of Ploughing

The number of ploughings necessary to obtain a good tilth depends on soil type, weed problem and crop residues on the soil surface. The heavy soil needs more number of ploughings, the range being 3-5 ploughings to obtain a good tilth. On the other hand, the light soils require 1–3 ploughings to obtain good tilth. More number of Ploughings are necessary for a soil having more weed growth and plant residues.

(ii) Secondary Tillage

The tillage operations that are performed on the soil after primary tillage to bring a good soil tilth are known as secondary tillage. Secondary tillage consist of lighter or finer operation which is done to clean the soil, break the clods and incorporate the manures and fertilisers. Harrowing and planking is done to serve those purposes. Planking is done to crush the hard clods, level the soil surface and to compact the soil lightly. Harrows, Cultivator, Spade, Ladder, Khurpi, Nirani *etc.* are used for secondary tillage.

(B) After Tillage

The tillage operations that are done in the standing crop after the growing or planting and prior to the harvesting of the crop plants are called after tillage. This is also known as intercultivation or post seeding/planting cultivation. It includes harrowing, hoeing, weeding, earthing up, drilling or side dressing of fertilisers *etc.* spade, nirani, wheel hoe, paddy weeder, *etc.* are used for intercultivation.

Tillage operations depend on the crops to be cultivated as follows :

(i) Dry Tillage

Dry tillage is generally practised for the crops that are sown or planted in dry land condition having sufficient moisture for germination of seeds. Dry tillage is suitable for the crops such as broad casted Paddy, Jute, Wheat, Oilseeds crop (*i.e.* mustard, sesame, linseed, sunflower *etc.*), Pulses (*i.e.* Gram, Black gram, Green gram, Lentil, Cowpea *etc.*), Potato, Radish, Carrot, Beet, Turnip and Vegetable crops. Dry tillage is done in a soil having sufficient moisture (21 – 23 per cent). The soil becomes more porous and soft due to dry tillage. Besides this, the water holding capacity of soil and aeration is also increased. This conditions are more favourable for soil micro-organism, the most important convertor of nutrients into available form to plants.

(ii) Wet or Puddling Tillage

The tillage operation that is done in a land with standing water is called wet or puddling tillage. Puddling operation consist of ploughing repeatedly in standing water untill the soil becomes soft and muddy. Puddling creates an impervious layer below the surface to reduce deep percolation, losses of water and to provide soft seed bed for planting rice. Puddling tillage is also done in both the direction for the incorporation of green manuring crops and weeds. Wet tillage destroys the soil structure and the soil particles that are separated during puddling settle later. Wet tillage is only the means of land preparation for transplanting semi–aquatic crop plant, such as rice *etc.* Planking after wet tillage makes the soil level and compact. Puddling hastens transplanting operation easily and smoothly as well as the establishment of seedlings. Wet land ploughs or worn-out dry land ploughs are generally used for wet or puddling tillage.

Factors Influencing Preparatory Tillage

The factors influencing preparatory tillage are as follows:

Crop

It dictates the type and extent of preparatory tillage. The small size seed crop (*i.e.* Mustard, Jute, Sesame, *etc.*) requires fine seed bed for good germination of seed and stand establishment. Pulse crop (*i.e.* gram, pea, pigeon pea, green gram *etc.*) does not require fine seed bed. The root crop (*i.e.* Radish, Carrot, Beet, Turnip *etc.*) and tuber crop (*i.e.* Potato *etc.*) require deep and fine tilth. Crops like Brinjal,

Cabbage, Cauliflower, Knolkhol, Chilli, Onion *etc.* are transplanted in the main field after raising the seedlings in the seed bed. The seed bed and main field require repeated ploughings followed by plankings for the sowing of seed and transplanting seedlings. Paddy require puddling condition which is achieved by tilling the land in standing water in the field.

(i) **Soil :** The light soil requires less tillage and heavy soil requires more tillage to bring the soil favourable for sowing seed and transplanting of seedlings.

(ii) **Climate :** The climatic condition affects on the moisture content of the soil which creates condition favourable for tillage.

(iii) **Cropping System :** Cropping system involving different crops need different types of tillage. Crops following pulse crops need lesser tillage than that of following maize, sugarcane *etc.* The crops following potato require minimum tillage. But the crops following paddy need repeated preparatory tillage for obtaining an ideal seed bed for sowing seed or transplanting seedlings.

(iv) **Type of Farming :** Tillage operation varies with the type of farming. Deep ploughing is necessary in dry land agriculture to eradicate perennial weeds as well as to conserve soil moisture. In intensive irrigated farming, tillage is required for each crop throughout the year.

Modern Concept of Tillage

Jethro Jull (1674–1741) gave a concept that 'Tillage is manure and this statement is correct in many sences'. Because tillage makes the soil a better medium for the growth and development of the plant. It improves the physical condition of the soil and makes better air-water temperature relationship in the soil. This condition helps in the better growth and activities of the soil microorganism. Tillage makes the soil suitable for sowing of seed and it also controls weeds in the field. In traditional tillage, too much tillage, is recommended to get fine tilth and it is expensive, time and energy consuming which often destroys the soil structure else. In recent time, the concept of tillage has been changed considerably and new concept have been introduced namely minimum tillage, zero tillage, stubble mulch tillage.

1. Minimum Tillage

Minimum tillage is aimed at reducing tillage to the minimum necessary for ensuring a good seed bed, rapid germination, a satisfactory stand and a favourable growing conditions. Tillage can be reduced into two ways such as (i) by emitting operations which do not give much benefit when compared to the cost and (ii) by combining agricultural operations like seeding and fertilizers application (Reddy and Reddi, 1993).

Minimum tillage can be practiced by different methods as follows :

(i) Row Zone Tillage : In this method, after completing the primary tillage in the entire field, the secondary tillage is done in row zone area only into which seeds are sown.

(ii) Plough Plant Tillage : In this method, after completing primary tillage in the entire field, the crop row area is pulverised and seeds are sown or seedlings are transplanted. In this system, ploughing and seeding or planting may be completed on one operation.

(iii) Wheel Tract Tillage : In this system, after completing the primary tillage as usual, tractor is used for sowing of seed and the wheel of tracted pulverises the soil of the crop row zone only.

Minimum tillage has some advantages and disadvantages as follows :

Advantages

(i) Improvement of soil condition due to decomposition of plant residues *in situ.*

(ii) Cost and time of soil preparation are reduced.

(iii) Higher infiltration caused by the vegetation presentation on the soil and channels formed by decomposition of seed, plant roots.

(iv) Less resistance to root growth due to improvement of soil structure.

(v) Less soil compaction by the reduced movement of heavy tillage implements.

(vi) Less soil erosion compared to conventional.

Disadvantages

(i) Seed germination is lower with minimum tillage.

(ii) Poor root and nodule development in some leguminous crops like peas and broad beans.

(iii) Reduced rate of decomposition of organic matter.

(iv) Necessity of herbicidal use to control weeds. Continuous use of herbicides causes pollution and dominance of perennial problematic woods.

Chapter 4

Crop Density and Geometry

Introduction

- ✰ Crop Density is just the number of plants within a given sample unit area or say per metre square.

- ✰ Cropping intensity refers to raising of a number of **crops in** the same field during one agriculture year. It can be expressed as. It is (Gross cropped area/Net sown area) x 100). For instance, if a farmer owns 5acre land and he takes two crops per year, one kharif and one rabi. He gets net effective produce from 10acres even though he owns 5acre of land.

- ✰ Crop geometry refers to the shape of the space available for individual plants. It influences crop yield through its influence on light interception, rooting pattern and moisture extraction pattern. Crop geometry is altered by changing inter and intra-row spacing (Planting pattern)

- ✰ Crop density in plant ecology is defined as the number of individuals of a given species that occurs within a given sample unit or study area. Number of plants per unit area in the cropped field is the plant population or plant density.

- ✰ Optimum plant population is the number of plants required to produce maximum output or biomass per unit area. Any increase beyond this stage results in either no increase or reduction in biomass.

Crop Density

Plant density is the number of individual plants present per unit of ground area. It is most easily interpreted in the case of monospecific stands, where all plants belong to the same species and have germinated at the same time. However, it could also indicate the number of individual plants found at a given location.

Right Planting Density

Planting densities that are either too low or too high may result in economic losses.

☆ *Impact of low planting density:* When the planting density is too low, each individual plant may perform at its maximum capacity, but there are not enough plants as a whole to reach the optimum yield. Therefore, total yield of the crop becomes a limiting factor. Also, when there is too much space left between plants, weed growth is promoted, which could increase weeding costs.

☆ *Impact of a high planting density:* If the planting density is too high, plants may compete against each other, known as intra-specific competition. Under those conditions, the performance of individual plants becomes a limiting factor for maximum crop yield.

☆ *Using the right density:* As the planting density increases, the total crop yield increases and reaches a maximum, at which point further increase in planting density results in reduced yield. It is, therefore, important to use appropriate planting density, keeping in mind the fact that the density that provides the highest yield and the density that provides the highest revenue may be different. Growers should shoot for both optimum yield and revenue when balancing their planting density.

Tips for Achieving the Right Planting Density

☆ Calibrate your seeder each time of use, especially if you change seed batches or the seeder has been sitting around for a long time or has been moved.

☆ Remember that for some crops seed size may vary from batch to batch and smaller seed could increase your planting density (unless you are using a planter with high accuracy in seed singulation).

☆ If you change your row spacing, remember to recalibrate the seeder for the target plant density. For example, reducing row spacing from 30 inches to 24 inches without recalibrating the seeder will increase the planting density by 25 percent.

☆ Maintain a uniform planting depth for uniform emergence.

Monostands

☆ Many of the processes related to plant density can well be studied in monocultures of even-aged individuals that are sown or planted at the same time. These can be referred to as 'monostands' and are often studied in the context of agricultural, horticultural or silvicultural questions. However, they are also highly relevant in ecology.

☆ In general, the total above-ground biomass of a monostand increases with increasing density, up to the point where the biomass saturates. This is what has been dubbed 'constant final yield', and refers to the total plant biomass per unit ground area.

☆ Seed production per ground area is not constant, but often declines with density after total biomass per ground area reached its maximum value.

Self Thinning and Plant Density

☆ Experiments with herbaceous plants have been carried out with extremely high densities (up to 80,000 plants per square meter). At such high densities, these plants will start to compete soon after germination, and eventually a large number of those individuals (up to 95 per cent) will die.

☆ In agriculture, farmers avoid these very high densities as they do not contribute to seed yield. Normal densities in modern agriculture depend on final plant size and vary between 5-10 plants per square meter for Maize till 200-300 plants per square meter for Rice or Barley. In forestry, normal densities are less than 0.1 plants per square meter. Not only the biomass per square meter increases with density, but also the Leaf Area Index (LAI, leaf area per ground area).

☆ The higher the Leaf Area Index, the higher the fraction of intercepted sunlight will be, but the gain in light interception and photosynthesis will not match the increase in LAI, and this is the reason that total biomass per ground area saturates at high plant densities.

Individual Plant in a Monostand

☆ **Biomass:** Contrary to the total biomass per unit ground area, which increases with density until reaching saturation, the average biomass of individual plants in a monostand strongly declines with plant density, such that for every doubling in density individual plants will become ~30-40 per cent smaller Plants in higher density stands invest relatively more of their biomass in stems (higher Stem Mass Fraction), and less in leaves and roots. Apart from their weight, plants will change their phenotype in many other ways and at different integration levels:[5]

☆ **Leaves:** Leaf size of the largest full-grown leaf of Maize plants grown at a low (L), intermediate (I), and high (H) plant density. Individual plants in dense stands have fewer leaves and they are often smaller and more narrow. Leaves of high-density plants are thinner (higher SLA – leaf area per unit mass), especially lower in the vegetation, with a similar concentration of nitrogen per unit mass, but a lower nitrogen content per area.

☆ **Stems:** Average plant height or vegetation height often remains remarkably similar, but a very consistent difference is that the stems of high-density plants have a much smaller diameter. They also have fewer side shoots (tillers) in the case of grasses, or branches in the case of herbs and trees.

☆ **Roots:** Root growth in environments with high plant density show that there will be fewer roots per plant and but the length and general density of the individual root remain somewhat the same, this is expected to still cause issues for the plant in future growth.

☆ **Physiology:** In dense stands, there is a strong gradient of light from top to bottom. Lower leaves in high-density stands will therefore have a lower photosynthetic rate and a lower transpiration rate than similar leaves of plants in open stands. There are indications that also the well-illuminated top leaves may have a lower photosynthetic capacity in densely-grown plants.

☆ **Seed production:** Because densely-grown plants are smaller, they will also produce fewer seeds per individual. But also the seed production as a fraction of total plant biomass (harvest index) is lower, and so is the seed weight of an individual seed.

Crop Geometry

☆ The arrangement of the plants in different rows and columns in an area to efficiently utilize the natural resources is called crop geometry. It is otherwise area occupied by a single plant *e.g.*, rice – 20 cm ×15 cm. This is very essential to utilize the resources like light, water, nutrient and space. Different geometries are available for crop production.

☆ The arrangement of the plants in different rows and columns in an area to utilize the natural resources efficiently is called crop geometry. It is otherwise area occupied by a single plant *e.g.*, Rice – 20 cm ×15 cm.

Types of Crop Geometry

☆ *Random plant geometry:* Random plant geometry results due to broadcasting method of sowing and no equal space is maintained. Resources are either underutilized or over exploited.

☆ *Square plant geometry:* The plants are sown at equal distances on either side. Mostly perennial crops, tree crops follow square method of cultivation. *e.g.*,. Coconut – 7.5 × 7.5 m; banana – 1.8 × 1.8 m. But, due to scientific invention, the square geometry concept is expanded to close spaced field crops like rice too.

- ☆ **Rectangular method of sowing:** There are rows and columns, the row spacing are wider than the spacing between plants. The different types exist in rectangular method are:

  - ❖ **Solid row**: Each row will have no proper spacing between the plants. This is followed only for annual crops which have tilling pattern. There is definite row arrangement but no column arrangement, Ex. Wheat.

  - ❖ **Skip row:** A row of planting is skipped and hence there is a reduction in population. This reduction is compensated by planting an intercrop; practiced in rain-fed or dry-land agriculture.

- ☆ **Triangular method of planting:** It is recommended for wide spaced crops like coconut, man *etc.* The number of plants per unit area is more in this system.

- ☆ **Quincunx or diamond pattern:** The quincunx or diamond pattern of arranging row-planted crops is a modified form of the square pattern. It consists of a square that is formed by 4 closest plants with an additional plant at the center of these 4 plants. The 4 plants that form a square are the main crops while the crop at the center is called a filler crop.

- ☆ **Miscellaneous planting arrangements:** Crops are sown with seed drills in two directions to accommodate more number of plants and mainly to reduce weed population. Crops like rice, finger millet are transplanted at the rate of 2-3 seedlings per hill. Transplanting is done either in rows or randomly. Skipping of every alternate row is skipped, and the population is adjusted by decreasing intra-row spacing, it is known as paired row planting. It is generally restored to introduce an intercrop.

Plant Population and Growth

- ☆ High plant density brings out certain modifications in the growth of plants. Plant height increases with increase in plant population due to competition for light.

- ☆ Sometimes it may happen that moderate increase in plant population may not increase but decrease plant height due to competition for water and nutrients but not for light.

- ☆ Leaf orientation is also altered due to population pressure. The leaves are erect narrow and are arranged at longer vertical intervals under high plant densities. This is a desirable architecture.

Plant Population and Yield

- ☆ Decrease in yield of individual plant at high plant density is due to the reduction in the no. or earls or panicles. Ex: - Redgram produces about 20 pods per plant at 3.33 lakh plants/ha (30×10cm) while it produces more than 100 pods per plant at 50,000 plants/ha (80×25cm).

☆ Under very high population levels plant become barren, hence optimum plant population is necessary to obtain maximum yield.

Optimum Plant Population

☆ Optimum plant population for any crop varies considerably due to environment under which it is grown. It is not possible to recommend a generalized plant population, since the crop is grown in different seasons with different management practices.

☆ *e.g.* Redgram plants sown as winter crop will have half the size of those grown in monsoon season. Optimum plant population is 55,000 plants/ha. For monsoon season crop of redgram and this is increased to 3.33 lakh plants/ha for winter crop; as low temperature retards the rate of growth, higher population is established for quicker ground cover.

Importance of Plant Population/Crop Geometry

☆ Yield of any crop depends on final plant population

☆ The plant population depends on germination percentage, and survival rate in the field.

☆ In rainfed conditions, high plant population will deplete the soil moisture before maturity, whereas low plant population will leave the soil moisture unutilized.

☆ When soil moisture and nutrients are not limited, high plant population is necessary to utilize the other growth factors like solar radiation efficiently.

☆ In low plant population individual plant yield will be more due to wide spacing.

☆ In high plant population, individual plant yield will be low due to narrow spacing leading to competition between plants.

☆ Yield per plant decreases gradually as plant population per unit area is increased, but yield per unit area increases up to certain level of population

☆ That level of plant population is called as optimum population

☆ So to get maximum yield per unit area, optimum plant population is necessary. So the optimum plant population for each crop should be identified.

Chapter 5

Manures and Fertilizers

Introduction

☆ The Manures is originated from the French Word "Manoeurver" which refers to "Work with Soil", that is why the word tillage and manures whose synonyms as it clear by the statement of Jethro Tull (1700 B.C.)— "Tillage is manure". The word of manure is also originated from Latin word *Manu* (hand) and *operare* (works).

☆ Manures are substances which are organic in nature, capable of supplying plant nutrient in available form, bulky in nature, having the low analytical value and having no definite composition and most of them are obtained from animal and plant waste products.

☆ Fertilizers are inorganic material which can supply

Manuring is the process of addition of natural or chemical sources of nutrients for the crop. Natural sources include dead wastes of plants, humans and animals, excreta and other wastes. These on decomposition give organic products called organic manure or simply manure. Manure improves the water holding capacity, aeration, and texture of the soil. These lead to the development of a new method of farming called organic farming where only organic fertilisers, pesticides, etc. are used.

Fertilisers are chemical compounds which include salts or organic compounds like urea, ammonium sulfate, sodium nitrate etc. They are the sources of plant nutrients like potassium, nitrogen, and phosphorus. Fertilisers are commercial products mainly manufactured in factories. They have enhanced the yield spontaneously.

plant nutrient in available form, having the high analytical value, and having a definite composition and mostly are industrial products. A fertilizers refers to a material added to the soil in order to supply chemical elements needed for plant nutrition and to improve soil fertility.

☆ Chemical fertilisers should be used in optimum amount with great care. Their excess use may lead to soil infertility, water pollution, and even cause disease. Overall, it damages the crop. Hence it is advisable to use organic manure instead of chemical fertilisers.

☆ Other alternative methods for soil replenishment are vermicomposting, crop rotation, growing of leguminous plants, *etc.*

Differences between Manure and Fertilizer

Manure	*Fertilizer*
Manure is an organic substance that is obtained from decomposition of vegetables and animal waste.	Fertilizers are inorganic substances manufactured in factories.
Manures are relatively less rich in plant nutrients, they only remove general deficiency of soil.	Fertilizers are very rich in plant nutrients like Nitrogen N, Phosphorus P, Potassium K.
They add humus to soil by providing organic substances and nutrients.	They ensure healthy growth and development of plants by providing nutrients however lack to add humus to soil.
The addition of manure to soil does not require any special guideline.	The addition of fertilizer to soil requires special guidelines like dose time and post addition precautions should be followed to avoid mineral toxicity.
Manure is not readily soluble in water thus it is absorbed by plants slowly.	Fertilizers are readily soluble in water thus are rapidly absorbed by plants.
It protects the environment and helps in recycling farm waste.	Its excessive use causes water pollution and cannot replenish organic matter of soil.

Role of Organic Matter in Soil Fertility

☆ Throughout history, people who raise livestock and poultry have used manure as a fertilizer, soil amendment, energy source, and even construction material. Manure contains many useful, recyclable components, including nutrients, organic matter, solids, energy, and fiber.

☆ Organic matter forms a very small but an important portion and it is obtained from dead plant roots, crop residues, various organic manures like farmyard manure, compost and green manure, fungi, bacteria, worms and insects.

☆ Organic matter improves the physical condition of the soil, particularly the structure.

❖ Decaying organic matter acts as a food material for bacteria, fungi and other organisms.

❖ Presence of organic matter dissolves many insoluble soil minerals and make them available to plants

❖ It plays an important role in the nutrient supplying power of soil as it has got high cation exchange capacity (CEC)

❖ It increases the water holding capacity of the soil, particularly in sandy soils.

❖ It improves aeration and infiltration in heavy soils.

❖ It reduces loss of soil by water and wind erosion

❖ It regulate soil temperature

❖ It serves as an important source of certain plant of food element (N, P, S *etc.*).

❖ The buffering nature of the organic matter is considered to be advantageous in the residue management of pesticides, herbicide and other heavy metals.

Classification of Manures and Fertilizers

Manures

I. Bulky Organic Manures

Compost

☆ Village compost

☆ Town compost, (made from farm waste products.

F.Y.M.

☆ Cattle manure

☆ Ship Penning

☆ Poultry manure

Sewage and Sludge

II. Green Manures

☆ Leguminous plant

☆ Non-leguminous plant

III. Oil Cake

Richest source of plant nutrient of all organic manures.

Edible Cake i.e., used for cattle feeding.

☆ Mustard cake

☆ Groundnut cake

☆ Linseed cake

☆ Sesame cake

Non edible cake i.e. used as manures

☆ Castor Cake

☆ Neem Cake

☆ Sunflower Cake

IV. Waste Product of Slaughter House

☆ Blood meal

☆ Bone meal

V. Fish Product

☆ Fish meal

VI. Guano

Material obtained from the excreta and dead bodies of sea bird.

Fertilizers

I. Nitrogenous Fertilisers

Sources of nitrogen.

☆ Fertilizer containing nitrogen in ammonium (NH_4) form: Such as Ammonium sulphate, Ammonium chloride *etc.*

☆ Fertilizer containing nitrogen in nitrate (NO_3) form : Such as Calcium nitrate, Sodium nitrate, Potassium nitrate *etc.*

☆ Organic fertilizer or fertilizer containing nitrogen in non–ionic form : Such as Urea, *etc.*

II. Phosphatic Fertilizers

Sources of Phosphorus:

☆ Fertilizers containing water soluble phosphorus : Such as Superphosphate, concentrated Superphosphate, Mono–ammonium Phosphate, Diammonium Phosphate, *etc.*

☆ Fertilizers containing citrate soluble Phosphate : Such as Dicalcium Phosphate, Basic Slag *etc.*

☆ Fertilizers containing insoluble phosphate. Such as Rock Phosphate, *etc.*

III. Potassic Fertilizers

Sources of Potassium : Such as Potassium Chloride, Potassium Sulphate *etc.*

IV. Compound Fertilizers

Sources of more than one nutrient : Such as Gromor (Urea Ammonium Phosphate), Mono–ammonium Phosphate, Diammonium Phosphate, Ammonium Phosphate, Suphala 20:20:0 (Ammonium Phosphate Sulphate), Suphala 15:15:15 (Nitro–Phosphate–Potash), IFFCO 10–26–26 *etc.*

V. Mixed Fertilizers

Source of three nutrients : Such as Grade–1 (8:8:8:) Fertiliser.

Manures

1. Farm Yard Manure (F.Y.M.)

It is a mixture of cattle dung, urine, litter or bedding material, portion of fodder not consumed by cattle and domestic wastes like ashes etc. collected and dumped into a pit or a heap in the corner of the backyard. It is allowed to remain there and rot till it is taken out and applied to fields. Well rotten Farm Yard Manure contains 0.5. per cent N., 0.2 per cent P_2O_5 and 0.5 per cent K_2O

☆ One of the most commonly used bulky organic manure is farm yard manure (F.Y.M.). F.Y.M. is produced from liquid and liquid excreta of cattle mixed with litter used for breeding purposes of cattle.

☆ The following method has been recommended by C.N. Acharyya for preparing good quality of F.Y.M. and to avoid high nutrient loss. F.Y.M. is prepared in trenches of 20 – 25 ft. (6 – 7.5 m) long, 5 – 6 ft. (150 – 180 cm) breadth and 3 – 3.5 ft (90 – 105 cm) deep. Farm waste mixed with earth should be under each animal in the evening for absorption of Urine. Each morning urine socked litter and dung should be mixed and taken to manure trench. A section of 3 ft. (90 cm) length is taken up from one end for filling with daily collection of refuses, when the trench is filled to a height of 1.5 – 2ft. (45 – 60cm) above ground level, the surface of the refuges is plastered with cow–dung mixed with soil. The manures becomes ready in about three months. By this time the second trench is being filled up.

☆ Generally two such trenches would be needed for 3–4 heads of cattle. It is possible to prepare by this process 250–300 cubic feet of manures (10–12 cartloads) per year per head of cattle.

2. Compost

☆ Compost is an another organic manure artificially prepared from vegetables and animal waste products. Two method of composting waste organic materials, such as aerobic and anaerobic are usually recommended.

☆ In the aerobic process, the mixed farm residues are collected in a heap and in the an–aerobic process, the mixed farm residues are collected in a pits of a convenient size, say 15' x 5' x 3' (450 x 150 x 90 cm). Each day a collection is spread in a thin layer and sprinkled with a mixture of fresh cow–dung (4.50 kg). Compost manure is reinforced the pits is filled to a height of 1.5 – 2 ft. (45 – 60 cm) above ground level, the surface is plastered with one inch layer of a mixture of mud and cow-dung. The compost becomes ready in about three to four months without any further attention.

Well rotted plant and animal residue is called compost. Composting means rotting of plant and animal remains before applying in fields. The essential requirements of composting are air, moisture, optimum temperature and a small quantity of nitrogen. It is an activity of micro-organisms and same people recommend addition of suitably prepared inoculums to introduce micro-organisms for decomposing the material.

Benefits of Composting

☆ Enriches soil, helping retain moisture and suppress plant diseases and pests.

☆ Reduces the need for chemical fertilizers.

☆ Encourages the production of beneficial bacteria and fungi that break down organic matter to create humus, a rich nutrient-filled material.

☆ Reduces methane emissions from landfills and lowers your carbon footprint.

Method of Compost Making

The main method of making compost is as follows:

☆ **Adco Process :** This process was introduced by Hutchinson and Richard in England in 1921. Adco–Powder, originated by Agricultural Development Company, England (Private concern operating at Herpenden, England) is used as a starter @ 7 kg. per 100 kg. of dry waste product.

☆ **Activated Compost Process :** This process was introduced by Fowler and Redge in 1922 at Indian Institute of Science, Bangalore. They used sewage sludge as a starter *i.e.* the material used for decomposing plant residues.

☆ **Indore Process :** This process was devised by Howard and Ward at the Institute of Plant Industry, Indore. The cow-dung in small amounts are used as 'Starter'. The Indore process is aerobic and decomposition takes place in aerobic condition.

☆ **Bangalore Process :** This process is devised by C.N. Acharya as a result of his experiment at the Institute of Science, Bangalore. Night Soil is used as a starter.

Note : 1 cft. refuge weighs roughly 20 lbs. 1 cft. of compost weighs roughly 40 lbs. One cart loads of refuse (30 cft.) weighs roughly 600 lb.

3. Compost Tea (Liquid Manure)

Compost tea is water in which compost has been steeped. Leached into that liquid are some of the compost's nutrients, microorganisms, and a witch's brew of poorly defined compounds called humates. Humates help plants better use nutrients already in the soil and offer a host of other benefits.

☆ Compost tea is prepared by decomposing the green leaves and leaves having pungent odour which are mixed with cow–dung or droppings of goat, sheep or poultry and taken in gunny bags which is kept in the drum having 200 litres of water.

☆ The compost tea is applied to the paddy in the morning or in afternoon. It can also be used as foliar spray.

☆ The optimum time of day to apply compost tea is in the morning, when plant stoma are open to receive it and the sun will dry leaves and prevent fungal diseases from excess moisture. Apply when soil is moist if using the product as a drench.

4. Vermicompost

☆ Vermicompost is now gaining popularity in Agriculture. It is made and used to recover their existence in the crop field. There is no existence of earthworm where chemical fertilizers and pesticides are used extensively. Vermi-compost is prepared in a pit specially by using earth worm. The available different kinds of farm waste are used for making vermi-compost. It is also available in the market.

☆ Vermicomposting is a type of composting in which certain species of earthworms are used to enhance the process of organic waste conversion and produce a better end-product. It is a mesophilic process utilizing microorganisms and earthworms. Earthworms feeds the organic waste materials and passes it through their digestive system and gives out in a granular form (cocoons) which is known as vermicompost.

☆ Vermicompost is earthworm excrement, called castings, which can improve biological, chemical, and physical properties of the soil. The chemical secretions in the earthworm's digestive tract help break down soil and organic matter, so the castings contain more nutrients that are immediately available to plants.

Applications of Vermicompost

☆ The worm castings contain higher percentage of both macro and micronutrients than the garden compost. Apart from other nutrients, a fine worm cast is rich in NPK which are in readily available form and are released within a month of application. Vermicompost enhances plant growth, suppresses disease in plants, increases porosity and microbial activity in soil, and improves water retention and aeration.

Vermicompost also benefits the environment by reducing the need for chemical fertilizers and decreasing the amount of waste going to landfills. Vermicompost production is trending up worldwide and it is finding increasing use especially in eastern countries, Asia-Pacific and Southeast Asia.

☆ A relatively new product from vermicomposting is vermicompost tea which is a liquid produced by extracting organic matter, microorganisms, and nutrients from vermicompost. Unlike vermicompost and compost, this liquid organic fertilizer may be applied directly to plant foliage, reportedly to enhance disease suppression. Vermicompost tea also may be applied to the soil as a supplement between compost applications to increase biological activity.

☆ Vermicompost may be sold in bulk or bagged with a variety of compost and soil blends. Markets include home improvement centers, nurseries, landscape contractors, greenhouses, garden supply stores, grocery chains, flower shops, discount houses, and the general public.

5. Green Manures

☆ The practice of turning into the soil undecomposed green plant tissue is referred to as green manuring (Buckman and Bredy, 1980) and the manures obtained by this method is known as green manure.

Green manuring is the practice of turning into the soil undecomposed green plant tissue. The function of a green manure crop is to add organic matter to the soil. As a result of the addition, the nitrogen supply of the soil may be increased and certain nutrients made more readily available, thereby increasing the productivity of the soil.

☆ Manures are plant and animal wastes that are used as sources of plant nutrients. They release nutrients after their decomposition.

☆ Manures are the organic materials derived from animal, human and plant residues which contain plant nutrients in complex organic forms.

Characteristics of Green Manure Crops

The green manure crop should possess some desirable characteristics :

☆ The green manure crops should have profuse leaves and rapid growth.

☆ The green manure crops should have abundance and succulent tops.

☆ The green manure crop should be able to grow well on poor soils.

☆ The green manure crop should have deep root system.

Other factors remaining the same, use of leguminous green manure crop is more useful in comparison to non–legumes as more nitrogen is added by legumes which will be advantageous for the soils and crops grown after green manuring.

Suitable Crops for Green Manuring : Crops suitable for green manuring are divided into two groups:

☆ **Non-Legumes :** The non–legumes used as green manuring crops provides only organic matter to the soil. *Examples* : Mustard, Wheat, Radish, Carrot *etc.*

☆ **Legumes :** The legumes used as green manuring crops provides nitrogen as well as organic matter to the soil. *Examples :* Sannhemp (*Crotolaria juncea*), Dhaincha (*Sesbania aculeata*), Senji (*Melilotus parviflora*) cowpea (*Vigna catjang*) *etc.*

Recently the leaves of *Glyricidia maculata* and *Sesbania speciosa* (Perennial and shrubs) are used as green manuring materials. These plants are planted in a bank of tank or fallow land and their leaves and succulent tops are added to the soil for green manuring.

Benefits of Green Manuring

There are numerous advantages of green manuring such as :

Green manures are plants which are grown to benefit the soil. Green manures are the organic way to improve the soil fertility, including adding valuable nitrogen, improve the soil structure, giving better drainage or water retention and suppress weeds attract beneficial insects and other predators.

☆ **Supplies of Organic Matter :** Green manures supply organic matter to the soil. Humus formed from green manures increases the adsorptive capacity of soil, promotes aeration, drainage, and granulation of soil particles which is helpful for plant growth.

☆ **Addition of Nitrogen :** The green manures crops supplies additional nitrogen in addition to organic matter, if it is a legume crop, which has the ability to fix nitrogen from the air with the help of nodule bacteria. So all the legumes crop leave the soil in better physical condition and richer in nitrogen content.

☆ **Nutrient and Soil Conservation :** Green manure crops act as cover crop. They prevent the soil erosion and nutrient loss by taking up soluble nutrients which might otherwise been lost in drainage water or due to erosion. Green manuring crop makes the availability of Phosphorus and other nutrients to the succeeding crops. Green manures have a marked residual effect also.

☆ *Increases the Biochemical Activity :* The organic matter added to soil by way of green manures acts as a food for microorganism and they stimulate bio-chemical changes.

Nutrient Content of Manures and Oil Cakes

Sl.No.	Manures	Nutrient Percentage		
		Nitrogen (N) (P_2O_5)	Phosphorus (K_2O)	Potassium
I.	**Bulky Organic Manure**			
	Cow dung (Fresh)	0.3–0.4	0.1–0.2	0.1–0.3
	Horse dung (")	0.4–0.5	0.3–0.4	0.3–0.4
	Sheep dung (")	0.5–0.7	0.4–0.6	0.3–1.0
	Night soil, Fresh	1.0–1.6	0.8–1.2	0.2–0.6
	Poultry manure, fresh	1.0–1.8	1.4–1.8	0.8–0.9
	Sewage, Sludge, dry	2.0–3.5	1.0–5.0	0.2–0.5
	Sewage, sludge, activated dry	4.0–7.0	2.1–4.2	.5–0.7
	Urine, Cattle	0.9–1.2	Trace	0.5–1.0
	Urine, Horse	1.2–1.5	Trace	1.3–1.5
	Urine, Human	0.6–1.0	0.1–0.2	0.2–0.3
	Urine Sheep	1.5–1.7	Trace	1.8–2.0
	F.Y.M.	0.5–1.5	0.4–00.8	0.5–1.9
	Compost (Town) dry	1.2–2.0	1.0	1.5
	Compost (Rural) dry	0.4–0.8	0.3–0.6	0.7–1.0
	Water hyacinth compost	2.0–3.0	1.0–2.0	3.0–4.0
	Green manure (Various, average)	0.5–0.7	0.1–0.2	0.6–0.8
	(*a*) Dhaincha	0.62	0.15	0.58
	(*b*) Cowpea	0.71	0.15	0.58
	(*c*) Sunn hemp	0.75	0.12	0.51
II.	**Manures of Animal Origin**			
	Dried blood	10.0–12.0	1.0–1.5	1.0
	Fish manure	4.0–10.0	3.0–9.0	0.3–1.5
	Bone meal (raw)	3.0–4.0	20.0–25	–
	Bone meal (steamed)	1.0–2.0	25.0–30.0	–
III.	**Wood Ash**			
	Ash, Coal	0.73	0.45	0.53
	Ash, Household	0.5–1.9	1.6–4.2	2.3–12.0
	Ash, Babul	0.1	0.8–1.3	1.5–3.1

Sl.No.	Manures	Nutrient Percentage		
		Nitrogen (N) (P_2O_5)	Phosphorus (K_2O)	Potassium
IV.	**Crop Residues**			
	Paddy husk	0.3–0.5	0.2–0.3	0.3–0.5
	Straw	0.36	0.08	0.71
	Groundnut shell and stem	1.6–1.8	0.3–0.5	1.1–1.7
V.	**Oil Cakes**			
	Edible oil cakes			
	Mustard cake	5.1–5.2	1.8–1.9	1.1–1.2
	Linseed cake	5.5–5.6	1.4–1.5	1.2–1.3
	Sesame cake	6.2–6.3	2.0–2.1	1.2–1.3
	Safflower cake (Decorticated)	7.9	2.2	1.9
	Groundnut cake	7.0–7.3	1.5–1.6	1.3–1.4
	Coconut cake	3.0–3.2	1.9–2.0	1.7–1.8
	Cotton seed cake (decorticated)	6.4	2.9	2.2
	Non-edible cake			
	Cotton seed cake (Undecorticated)	3.9–4.0	1.8–1.9	1.6–1.7
	Castor cake	4.3	1.8	1.3
	Karanj or honge cake	3.9	0.9	1.2
	Mahua cake	2.5–2.6	0.8–0.9	1.8
	Neem cake	5.2–5.3	1.0–1.1	1.4–1.5

6. Oil Cakes

Oil cakes are the by-products of oilseed crops. Oil cakes are the important and quick acting organic nitrogenous manures.

It also contains small amount of phosphorus and potassium. A large variety of oil cakes are produced in the country.

They can be grouped into two classes as follows :

Edible Oil Cakes

This types of oil cakes are mainly used as cattle feed. But it can also be used as organic manure in the field.

Examples : Mustard cakes, groundnut cakes, Sesame or til cakes, Linseed cakes, coconut cake *etc.*

Non-Edible Oil Cakes

This types of oil cakes are not suitable for feeding cattle and mainly used for manuring crops.

Examples : Castor cake, Neem cake, Karanj cake, Mahua cake, Safflower cake *etc.*

☆ Oil cakes are usually used as organic nitrogenous manures. They can be applied as basal dressing (*i.e.* a few days prior to rowing, at sowing time of the crop) and as a top dressing after the crop has made a certain amount of growth. Oil cakes should be well powdered before application so that the manure is spread uniformly (Daji, 1955).

7. Blood Meal

☆ Blood meal is obtained from slaughter house. It is a very quick acting manures and is effective on all crops and on all types of soil.

☆ Blood meal is a nitrogen amendment that you can add to your garden. Adding blood meal to garden soil will help raise the level of nitrogen and will help plants to grow more lush and green.

☆ The nitrogen in blood meal can also help raise the acid level of your soil, which is beneficial to some kinds of plants that prefer soils with low pH (acidic soil).

☆ Blood meal is a organic manure of animal origin. It is used as manure and also used as ingredient of poultry feed. It should be applied to crops like oil cakes.

8. Fish Manures

☆ Non-edible fish carcasses of fish and offal are used to prepare fish meal. It is available either as dried fish or fish meal or powder. It contains 4–10 per cent organic nitrogen, 3–9 per cent phosphorus and 0.3–1.5 per cent potassium.

☆ As fish fertilizer improves soil health, it also increases soil fertility by providing the primary nutrients necessary for plants to thrive. Fish fertilizers offer a source of burn-free nitrogen, along with the other primary nutrients of phosphorus and potassium.

☆ Fish manure is a quick acting organic manure. It is suitable for application to all crops and on all soils. It is also used as an ingredient of poultry feed.

9. Wood Ash

☆ Wood ash, Cattle dung ash *etc.* are the indigenous source of potassium. Wood ash is a good source of potassium. But it also contains Nitrogen and Phosphorus.

☆ Ash is also a good source of potassium, phosphorus, and magnesium. In terms of commercial fertilizer, average wood ash would be about 0-1-3 (N-P-K). In addition to these macro-nutrients, wood ash is a good source of many micronutrients needed in trace amounts for adequate plant growth.

☆ Wood ashes are applied in all crops. It is a good manure for onion, garlic *etc.* Cowdung ashes alone or mixing with kerosine oil are dusted on vegetable crops as repellent for the insects.

10. Sewage and Sludge

☆ Sewage and sludge is the product of sewage system of sanitation, a modern systems of sanitation adopted in cities.

☆ In general, sewage has two components as follows :

❖ Solid portion, technically known as sludge.

❖ Liquid portion commonly known as sewage water. It is known as treated effluents.

❖ The sludge is collected separately, dried and used as manure.

☆ Sludge can profitably used as organic manure for producing crop. Sewage water can be used for irrigation. Both sewage and sludge are used in the compost making by activated compost process.

11. Poultry Manure

☆ The excreta of bird and manures obtained from deep litter system of poultry farming are the good source of organic manures.

Poultry manure contains all 13 of the essential plant nutrients that are used by plants. These include nitrogen (N), phosphorus (P), potassium (K), calcium (Ca), magnesium (Mg), sulfur (S), manganese (Mn), copper (Cu), zinc (Zn), chlorine (Cl), boron (B), iron (Fe), and molybdenum (Mo).

☆ Farmers with farms close to poultry farms use poultry manure regularly for their crops, with good returns. Poultry manure is a more concentrated source of crop nutrients, especially NPK and calcium. Being naturally organic, it does not need composting and can be applied directly to the fields from the farm.

☆ Poultry manures are a good organic manure for all crops. It should be applied as basal dressing.

☆ Composted chicken manure provides a slow-release source of macro- and micronutrients and acts as a soil amendment. Compared to other manures, chicken manure and the associated litter are higher in nitrogen, potassium, phosphorus and calcium, and are also rich in organic matter (Zublena, 1993).

12. Night Soil

Night soil is the human excrement. Dehydration of night soil can be done as such or admixture of absorbing materials such as soil, ash, charcoal and saw dust. Saw dust has high dehydration capacity as well as property of absorbing fuel smell.

Uses of Night Soil

Night soil is a good manure of producing crop. But there is some prejudice of using night soil for crop production. Night soil is used as starter in compost making by Bangalore process.

13. Biofertilizers

The term "Biofertilizers" includes selective micro organism like algae, fungi and bacteria. This microorganisms are capable of atmospheric Nitrogen or converting insoluble phosphorus into soluble form. Biofertilizers are ecofriendly and renewable sources of plant nutrients to supplement chemical fertilizers.

Classification of Biofertilizers

The biofertilizers can be classified depending on the types of microorganisms as follows:

☆ *Algal Biofertilizers:* An Algae Biofertilizer is a natural, organic and renewable energy source. They help retain essential nutrients and water in the soil which is required for the proper growth of the plants. *Example:* Azolla, Blue Green Algae (BGA) *etc.*

☆ *Bacterial Biofertilizers:* Plant growth-promoting rhizobacteria (PGPR) are naturally-occurring soil bacteria able to benefit plants by improving their productivity and immunity. These bacteria are associated with the rhizosphere, the part of soil under the influence of plant roots and their exudates. *Example :* Azospirillum, Azotobacteria, Phosphobacteria (*i.e.* Phosphorus solubilizing bacteria), Rhizobium *etc.*

☆ *Fungal Biofertilizers:* Fungal biofertilizers which involve the use of fungal agents (*Mycorrhiza* sp., *Trichoderma* sp., *Chaetomium* sp., and *Gliocladium* sp.) are formulated to provide nutrients to the host plant and safeguard crops against pathogens. *Example :* Mycorrhiza *etc.*

☆ *Actinomycetes Biofertilizers:* The use of actinomycetes for improving soil fertility and plant production is an attractive strategy for developing sustainable agricultural systems due to their effectiveness, eco-friendliness, and low production cost. *Example :* Frankia *etc.*

Uses of Biofertilizers

Biofertilizers play a vital role in maintaining long term soil fertility and sustainability. Biofertilizers are used in different ways in the production of field crops as follows :

☆ *Seed Treatment or Seed Innoculation :* Biofertilizers (*i.e.* Azotobacter, Azospirillum, Rhizobium, Phosphobacteria *etc.*) are used for seed treatment.

☆ *Dipping of Seedlings:* The seedlings are dipped in the solution of biofertilizers before transplanting. The treated seedlings are then transplanted in the field. Azospirillum is used for dipping of paddy seedlings.

★ **Soil Application:** The biofertilizers are mixed with well dried, powdered F.Y.M. (Farm yard manure) or compost and then broadcasted in the field just before sowing of seed or transplanting of seedlings.

Fertilizers

I. Nitrogenous Fertilizers

1. Ammonium Sulphate $(NH_4)_2SO_4$

It is most widely used fertilizers in our country. It is white crystalline salt and contain 20.5–21 per cent ammoniacal nitrogen. It is quick acting fertilizer with good storage quality. Ammonium sulphate can be applied either at sowing time or as a top dressing *i.e.* during growing period of the crop. It should not be applied along with the seed. It is a suitable fertilizer for all crops and for a wide variety of soils. Ammonium sulphate physiologically is acidic in nature. Acid equivalent is 107 kg of Calcium Carbonate per unit of nitrogen *i.e.* 107 kg Calcium Carbonate is required to neutralize the acidity produce by application of 20 kg. nitrogen as Ammonium sulphate.

2. Ureas $Co(NH_2)_2$

Urea is a white, crystalline product, soluble in water, hygroscopic and it has a tendency of caking. It is highly concentrated nitrogenous fertilizers containing 44–46 per cent organic nitrogen. It is very soluble in water and therefore, subject to rapid leaching. It may be applied at sowing time or as a top dressing. When applied at sowing time, it should not be allowed to come in contact with seed. Now–a–days Urea is being applied as foliar spray. It is suitable for application to all crops and soil when the soil is not wet. Physiologically Urea is a acid forming fertilizer and its acid equivalent is 36 kg $CaCO_3$ per unit of nitrogen.

3. Ammonium Chloride (NH_4Cl)

It is a white crystalline substances and contain 26 per cent nitrogen. It is very much soluble than ammonium sulphate. In India, it is largely used for industrial purposes. Its behaviour is same as Ammonium Sulphate.

4. Sodium Nitrate $(NaNO_3)$

It is a white crystalline substances and very hygroscopic in nature. It contain 16 per cent nitrogen. It is not suitable fertilizer for waterlogged condition, because large portion of fertilizer is lost by leaching. It is a good for wheat, maize, barley, cotton, sugarcane *etc.* It is physiologically basic in nature and its basic equivalent is 27 B lb. or calcium carbonate per unit of nitrogen.

5. **Ammonium Nitrate (NH_4NO_3)**

It is a white crystalline salt, very soluble in water, and hygroscopic in nature. It contains 33–35 per cent nitrogen, half as nitrate nitrogen and half in the ammonium form. It is also a acid forming fertilizers.

6. **Ammonium Sulphate Nitrate [(NH_4)$2SO_4$ + NH_4NO_3)]**

It is a mixture of ammonium nitrate and ammonium sulphate. It is available in white Crystalline form or as granular form of a dirty white colour. It contains 26 per cent nitrogen, three forth being in ammoniacal form and the rest (6.5 per cent) in nitrate form. It is soluble in water and is readily available to crops. It can be applied before sowing, at sowing time or as top dressing. It is useful for all crops and is suitable for application to all types of Soil. It also produces acidity in soil but to a much less extent (about half) than ammonium sulphate.

7. **Calcium Cyanamide ($CaCN_2$)**

It is a dark grey powdered substances. It is not hygroscopic and therefore easy to store. It is a poisonous substances. So care should be taken during handling. It contain 20–21 per cent nitrogen. The basicity is 57 lb of calcium carbonate per unit of nitrogen.

II. Phosphatic Fertilizers

1. **Superphosphate**

   ☆ It is the most widely used phosphatic fertilizer in India. It is a grey ash like powder with good keeping quality. It is manufactured largely by treating ground phosphate rock with almost an equal quantity by weight of sulphuric acid.

   ☆ The fertilizer is manufactured in three grades : *viz.* Single, double and triple superphosphate. At present only single superphos phate is available in the market. Double and triple superphos phate is prepared by reaching rock phosphate with a mixture of phosphoric acid and sulphuric acid respectively. The phosphoric acid in the superphosphate is in the water soluble form which is readily available for nutrition of plants. Although superphosphate readily dissolved in water, it does not wash out from the soil.

   ☆ It is suitable for all crops and can be applied to all soil except acid soil. In acid soil, it is immediately converted into insoluble phosphate due to precipitation as iron and aluminium phosphate. It should be applied before or at the time of sowing.

2. **Bone Meal**

   ☆ It is both phosphatic manures and fertilizer. It contains a small quantity of nitrogen also. It is available in two forms. (*i*) raw bone meal and (*ii*) steamed bone meal.

☆ Bone meal having particles not large than 3/32 inch in size is considered suitable for the use as fertilizers. Bone meal may be applied to the soil either at sowing time or just before it.

☆ It is a most suitable fertilizers for acid soil and is considered a safe manure for all crops.

III. Potassic Fertilizers

1. Muriate of Potash or Potassium Chloride (KCl)

It is a white or reddish to white crystalline or powdered materials. It contains 50–60 per cent potash, the whole of which is readily available. It is soluble in water but it is not lost from the soil as it is absorbed by the soil particles. It can be applied at sowing time or prior to sowing. Now a days, it is being applied as top dressing.

Nutrient Contents of different Fertilizers (In Percentage)

Name of Fertilizers	*1	2	3	4	5	6	7
I. Nitrogenous Fertilizers							
Ammonium Sulphate	20.6	–	–	–	–	110	–
Ammonium Chloride	25.0	–	–	–	–	148	–
Ammonium Sulphate nitrate	26.0	–	–	–	–	93	–
Calcium Ammonium nitrate	25.0	–	–	10-20	7-7.5	–	–
Ammonia anhydrous	82.0	–	–	–	–	148	–
Sodium Nitrate	16.0	–	–	–	–	80	–
Urea	44-46	–	–	–	–	80	–
II. Phosphatic Fertilizers							
Super Phosphate (Single)	–	16-20	–	25-30	0.5	–	–
Super Phosphate (Double)	–	35-38	–	–	–	–	–
Super Phosphate (Triple)	–	46-50	–	17-20	0.5	–	–
Basic Slag	3.0-8.0	–	–	–	–	–	–
Bone meal (raw)	3.0-4.0	20-25	–	–	–	–	–
Bone meal (steamed)	1.0-2.0	25-30	–	–	–	–	–
Rock Phosphate							
(*i*) Mussouri Phos	–	23.0-24.0	–	–	–	–	–
(*ii*) Purulia Phos		23.0					
III. Compound and Complex Fertilizer							
Monoammonium Phosphate	11.0	48.0	–	–	–	–	–
Diammonium Phosphate (D.A.P.)	18.0	46.0	–	–	–	–	–

Name of Fertilizers	*1	2	3	4	5	6	7
Ammonium Phosphate (Paramphosh)	16.0	20.0	–	–	–	–	–
Suphala 20:10:0	20.0	20.0	–	–	–	–	–
Suphala 15:15:15 (Nitrophosphate potash)	15.00	15.0	15.0	–	–	–	–
Gromor 28:28:0	28.0	28.0	–	–	–	–	–
(Urea ammonium Phosphate)							
IFFCO 10:26:26	10.0	26.0	26.0	–	–	–	–

*1: Nitrogen; 2: Phosphorus; 3: Potassium; 4: Lime (Cao); 5: MgO; 6: Equivalent Acidity; 7: Equivalent Basicity.

2. Potassium Sulphate or Sulphate of Potash K_2SO_4

It is a another potassic fertilizer made by treating potassium chloride with magnesium sulphate. It contain 48 to 52 per cent potash. It dissolves readily in water and becomes available to the crop almost immediately after application. It can be applied at any time upto sowing. It is considered better than Muriate of potash for crops the tobacco, fruit trees and potato *etc.*

Mixed Fertilizer

Mixed fertilizer is one which contain two or three fertilizers and available in the market in a particular grade (*e.g.* Grade – 1–8: 8: 8). The fertilizer grade refers to the minimum guarantees of plant nutrient content in terms of total nitrogen, available phosphoric acid and water soluble potassium. We can make the fertilizer mixture by adopting the following formula :

Calculation

Amount of nutrient in mixed fertilizer × Amount of mixture to be prepared

_____

Amount of nutrient in the ingredient.

One fertilizer can not be mixed with other fertilizer. Because uneven mixing of incompatible fertilizers leads to a loss of some of the fertilizer nutrients in the form of gas, convert soluble nutrients into insoluble ones or produces caking.

Certain fundamental principles are to be followed in mixing fertilizer such as :

☆ Ammonium Sulphate, Ammonium Chloride and other ammoniacal fertilizers and nitrogenous organic manures should not be mixed with lime.

☆ Calcium cyanamide, Basic slag, quick lime, slaked lime should not be mixed with fertilizers containing nitrogen in ammoniacal form.

☆ Urea should not be mixed with superphosphate and superphosphate should not be mixed with lime or Calcium Carbonates or wood ashes.

☆ Sodium nitrate or potassium nitrate should not be mixed with superphosphate.

☆ Ammonium sulphate nitrate should not mixed with lime.

Methods of Fertilizer Application

The fertilizers are generally applied to crop by different methods as follows :

Application of Fertilizers in Solid Form

1. Broad Casting

The fertilizers are spread by hand in the soil with last preparatory tillage just before sowing of seed or planting of seedlings. This method is known as "broadcast application" there are two types of broadcasting method of fertilizer application depending on the time of application as follows :

☆ *Broad Casting at Planting :* The fertilizers are broad casted to soil just before planting. The applied fertilizers are incorporated to soil by ploughing followed by planting.

☆ *Top-Dressing :* The method of application of fertilizer in standing crop is known as 'Top–dressing' generally nitrogenous and potassic fertilizers are applied to crop as top dressing.

2. Placement

It is a method of placing fertilizer in the soil before sowing or after sowing the crops. The followings are the common methods of this category.

☆ *Plough Sole Placement :* The fertilizers are placed in the plough sole after opening the furrow by plough and these furrows are covered immediately as the next furrow it is turned.

☆ *Deep Placement :* The nitrogenous fertilizers is placed deep in reduced layer to check denitrification.

☆ *Sub-Soil Placement :* The phosphatic and potassic fertilizers are placed in the sub-soil with the help of heavy machinery to avoid their fixation in strongly acidic soil.

3. Localised Placement

It is a method of placing fertilizer into the soil close to seed or plant. The roots of young plant can get nutrition as per their requirement from the fertilizer applied by this method. This method is usually employed when relatively small quantity of fertilizer is to be applied. Localised placement reduces fixation of phosphorus and potassium (Yawalker *et al.*). It is two types as follows :

☆ **Contact Placement or Combined Drilling or Drill Placement :** Contact placement refers to the drilling of seed and fertilizer together during sowing. The phosphatic and potassic fertilizer in small quantities are applied to cereal crops and cotton. But this method is not suitable for pulse crop. Seed-cum-fertilizer drill is used for such placement. In the absence of seed-cum-fertilizer drill, small doses of granular fertilizer may be mixed with seed and drilled together. However where using higher dose of fertilizer, adoption of this practice, even under irrigated condition, will not advisable.

☆ **Band Placement :** In this method the fertilizers are placed in bands on one side or both sides of the row at about 5cm away from the seed or plant in any direction.

☆ **Row Placement :** The fertilizers are placed on one or both sides of the row in continuous band by hand or seed drill. This method is practised in row crops (*e.g.* Sugarcane, Potato Tobacco, Cotton and Vegetables) with wide space between rows. This method is suitable for applying large quantities of fertilizers.

☆ **Side Dressing :** In this method, fertilizers are applied along the side of row or around the plant and mixed them with the soil by Spade or Khurpi or Nirani. It is of two types as follows:

❖ **Spot Placement :** The fertilizers are applied in spaces in between the plants and mixed them with soils by Khurpi or Nirani. This method is employed for applying fertilizer in vegetable crops.

❖ **Basin Placement :** The fertilizers are applied in the shallow basin already made around the base of the fruit trees. The applied fertilizers are mixed with the soil by Spade, Khurpi or Nirani.

4.   Pellet Application

Small pellets of convenient size are made after mixing the nitrogenous fertilizers specially urea with the soil and they are applied one to two inches deep between the rows of paddy crop. The pellets are deposited in soft mud of paddy field. This method of fertilizer application decreases the nitrogen loss through leaching or by runoff of water. Paddy response well to pellet application of Nitrogenous fertilizers compared to top dressing which is the most common method followed by cultivators. This is the findings of field experiment at C.R.R.I., Cuttack.

Application of Fertilizers in Liquid Form

Liquid form of fertilizers are applied by the following methods :

1.   Starter Solution

Starter Solution is prepared by mixing N.P. and K. fertilizer in the ratio of 1:2:1 or 1:1:2. This is applied to vegetable seedlings at the time of transplanting and it helps rapid establishment and quicks early growth of seedlings.

2. Foliar Application

This is a method of spraying on leaves of growing plant with suitable fertilizer solution having low concentration. Application of one fourth to one third of Nitrogen through foliage has been found to increase wheat and rapeseed yield to the extent of 10–15 per cent under unirrigated condition and over drilling the entire dose of nitrogenous fertilizer at sowing of seed. Deep placement of super granule urea (S.G.W.) resulted in one third saving in Nitrogen dose in Paddy over broadcasting urea of ordinary grade (Sharma *et al.*, 1980).

There are certain difficulties associated with foliar application of nutrients as enumerated by Yawalker *et al.,* 1977. These are as follows :

☆ Marginal leaf burn or scorching may occur if strong solution is used.

☆ As solution of low concentration (usually 3–6 per cent) is to be used, only small quantities of nutrients can be applied in one single spray.

☆ Several applications are needed for moderate to high fertilizer rates.

☆ Foliar application of fertilizer is costly compared to soil application, unless combined with other spraying operation taken up weed, insect or disease control.

3. Direct Application to the Soils

Liquid fertilizers such as Anhydrous Ammonia and Nitrogen solution are directly applied to the soil with the special injection equipment. Liquid manures such as urine, sewage water and cowshed washing are let into the field (De, 1990).

4. Fertigation

Fertilizers, both straight and mixed fertilizers are easily soluble in water. They are allowed to dissolve in irrigation water and applied to the soil through irrigation water. Nitrogenous fertilizers are generally applied through irrigation water.

Tips To Get Best Efficiency of Applied Fertilizers

☆ The fertilizer scheduling must be based on soil test.

☆ Selection of fertilizers should be done according to the soil reaction *viz.* acidic fertilizer for alkaline soils and basic fertilizers for acidic soils.

☆ Surface application through broadcasting should not be adopted, but fertilizer should be placed about 3–4 cm by the side or below the seed. This discourages weed growth also.

☆ The phosphorus and potassic fertilizers should be basal dressed, because P_2O_5 and K_2O are not lost from the soil but they are absorbed by soil particles. Their poor mobility restricts them to the place of application, therefore, they must be placed in the root zone.

☆ Home mixing of fertilizers should be in accordance with fertilizer mixing guide and such fertilizer mixture must be applied as soon as possible.

☆ In case of heavy soil type, half of nitrogenous fertilizers should be basal placed and rest half should be top dressed in one split only, but in case of light soil, Nitrogen should be applied in three equal splits *i.e.* 1/3 as basal dressing, 1/3 after 30 days of sowing and rest 1/3 about 50–60 days after sowing.

☆ Flooding with too deep water or poor drainage should be avoided after application of fertilizers at least for a week time.

☆ Top dressing should be done after draining out the water and weeding so that loss of nutrient's is minimum.

☆ The Paddy field, used for transplanting should be puddled, and fertilizers should be applied at the time of puddling, because this will help the fertilizers to reach and get store in reduced zone of the soil.

☆ Light sandy, calcarious and soils under very high cropping intensity become deficient in micronutrients like zinc and sulphur. The deficient plants become sicky, and cannot absorb nutrients, thus the fertilizer is not absorbed, therefore such soils must be supplied with Zinc sulphate ($ZnSO_4$) at the rate of 10–15 kg/ha 2–3 years.

☆ The acidic soils should treated with liming materials as and when required.

☆ Deep placement of fertilizer along with foliar feeding of Nitrogen (*i.e.* urea) through spraying of Nitrogenous fertilizer in place of top dressing should be done at least once in 3–5 years of time.

☆ Weed growth should not be permitted in cropped area during any part of the year.

☆ In case of flooded soil or calcarous soils, used of slow release Nitrogenous fertilizers like U.F.-30, Sulphur coated urea, Super granules, Neem coated or neem blended urea should be done, so that loss of Nitrogen can be minimised.

☆ Mud bolls contains urea, should be used in case of deep water crops, because it helps in proper placement and also reduces the loss of Nitrogen from the field.

☆ An appropriate plant protection measures and proper tillage practices should be adopted, so that plants remain healthy and absorb the applied nutrients from the field. (*Source* : Principles and Practices of Agronomy (1988), by S.S. Singh (p. 81–82).

Chapter 6
Essential Plant Nutrients

Introduction

Nutrient: A nutrient element is one that is required to complete the life cycle of the organism and its relative deficiency produces specific deficiency symptoms. Nutrients content is considered deficient when it is so low that it severely limits growth and produces characteristic deficiency symptoms. Range of nutrient content in plants associated with optimum crop yields is called sufficient. When the concentration of a nutrient element rises too high to cause significant growth reductions, it is termed as toxic.

Available nutrients: In the soil a nutrients elements is distributed in different discrete chemical forms, which often exist in a state of dynamic equilibrium and constitute the pool from which plants draw it. The nutrient available to the biological organism is termed as bio available nutrients. It is that portion of the nutrient in the soil that can be readily absorbed and assimilated by the plants. An available nutrient constitutes only a small portion of the total nutrient present in the soil.

Beneficial elements: Beneficial elements are the mineral elements which stimulate plant growth but are not essential or which are essential only for certain plant species, or under specific conditions like silicon, sodium, aluminium, cobalt, selenium and vanadium.

Fractional nutrient: This term introduced by Nicholas (1961) is defined as an element that plays a role in plant metabolism, whether or not that role is specific or indispensable.

Trace element: Trace elements are an element found in low concentration, perhaps less than one ppm or still less in soil plant and water, *etc.*

Tracer element: Radioisotope or a stable isotope of an element used for tracing its path in a system to study the mechanism of its interaction with the system is called a tracer element.

Heavy metal: A metal having specific gravity of more than 5.0 or having atomic number higher than 20 is termed as a heavy metal. As a corollary, any metal heavier than calcium is a heavier metal.

Nutrient content: Concentration of a nutrient or its amount per unit weight of a plant tissue is termed as nutrient content. Nutrient content is expressed in term of percent (kg/100kg or g/100 g) or ppm (parts per million) which is equivalent to mg of nutrient/kg of dry matter (mg/kg) or μg of nutrient per g of dry matter (μg/g). The percent (per cent) of a nutrient is converted to parts per million by multiplying with a factor of 10,000 or 10^4.

Nutrient Accumulation: storage of a nutrient in a particular part or portion of the plant is called nutrient accumulation.

Nutrient uptake: Amount of nutrient taken up by the growing crops from either the soil or other sources is called nutrient uptake.

Nutrient Removal: The nutrient contained in the harvested portion of the crop is termed as the nutrient removed.

Essential Plant Nutrients

☆ Plants require 17 essential elements for growth: carbon (C), hydrogen (H), oxygen (O), nitrogen (N), phosphorus (P), potassium (K), sulfur (S), calcium (Ca), magnesium (Mg), boron (B), chlorine (Cl), copper (Cu), iron (Fe), manganese (Mn), molybdenum (Mo), nickel (Ni), and zinc (Zn). These 17 essential elements, also called nutrients, are often split into three groups. The first group is the three macronutrients that plants can obtain from water, air, or both— carbon (C), hydrogen (H) and oxygen (O). The soil does not need to provide these nutrients, so they are not sold as fertilizers.

17 are essential nutrients. In the absence of one essential nutrient, the presence of all other essential nutrients becomes meaningless because the plants will not be able to utilize these nutrients for its growth and development. The essential nutrient elements are carbon, hydrogen, oxygen, nitrogen, phosphorus, potassium, calcium, magnesium, sulphur, iron; manganese, zinc, copper, boron molybdenum, chlorine, and nickel. These are recognized as universally essential nutrient elements, because these meet the essential requirements for diverse groups of organism algae bacteria fungi and green plants. Cobalt has been established to be essential for leguminous plants only.

☆ The other 14 essential elements are split into two groups—soil-derived macronutrients and soil derived micronutrients. This split is based on the actual amount of nutrient required for adequate plant growth. The soil-derived macronutrients are nitrogen (N), phosphorus (P), potassium (K), sulfur (S), calcium (Ca), and magnesium (Mg). The soil derived micronutrients are boron (B), chlorine (Cl), copper (Cu), iron (Fe), manganese (Mn), molybdenum (Mo), nickel (Ni), and zinc (Zn).

Soil-Derived Macronutrients

The six soil-derived macronutrients are present in plants at relatively high concentrations—normally exceeding 0.1 percent of a plant's total dry weight. This translates into a minimum need of 20 pounds of each macronutrient per acre each year.

Uptake Form and Typical Plant Content of the
14 Soil Derived Essential Nutrients

Essential Nutrient	Uptake Form	Plant Content (Dry Weight)	
		Average	Range
		(Per cent)	
Nitrogen	NO_3^-, NH_4^+	1.5	0.5-5.0
Phosphorus	$H_2PO_4^-, HPO_4^{2-}, PO_4^{3-}$	0.2	0.1-0.5
Potassium	K^+	1.0	0.5-5.0
Sulfur	SO_4^{2-}	0.1	0.05-0.5
Calcium	Ca^{2+}	0.5	0.5-5.0
Magnesium	Mg^{2+}	0.2	0.1-1.0
		ppm	
Boron	$H_2BO_3^-, H_2BO_3^-, HBO_3^{2-}$	20	2-100
Chlorine	Cl^-	100	80-10.000
Copper	Cu^{2+}	6	2.20
Iron	Fe^{2-}	100	50-1,000
Manganese	Mn^{2+}	50	20-200
Molybdenum	MoO_4^{2-}	0.1	0.05-10
Nickel	Ni^+	<<<0.0001	?
Zinc	Zn^{2+}	20	10-100

Table 2: Function and Mobility within Plant Tissue of the
14 Soil Derived Essential Nutrients for Plant Growth

Essential Nutrient	Mobility in Plant	Function of Plant
Nitrogen	Good	Proteins, protoplasts, enzymes
Phosphorus	Good	ATP, ADP, basal metabolism
Potassium	Good	Water relations, energy relations, cold hardiness

Essential Nutrient	Mobility in Plant	Function of Plant
Sulfur	Fair/Good	Proteins, protoplasts, enzymes
Calcium	Very poor	Cell structure, cell division, cell elongation
Magnesium	Good	Chlorophyll, enzymes
Boron	Very poor	Sugar translocation, cell development, growth regulators
Chlorine	Good	Photosynthesis
Copper	Poor	Enzyme activation
Iron	Poor	Chlorophyll synthesis, metabolism, enzyme activation
Manganese	Poor	Hill reaction-photosystem II, enzyme activation
Molybdenum	Poor	Nitrogen fixation, nitrogen use
Nickel	Unknown	Iron metabolism
Zinc	Poor	Protein breakdown, enzyme activation

Functions and Deficiency Systems of Essential Plant Nutrients

☆ Plants require 17 essential nutrients for their normal growth and completion of life cycle.

☆ On the basis of their relative concentration or amount in plant tissue, these are divided into (i) macronutrients and (ii) micronutrients

Macronutrients

☆ These are required in large amount (g/kg dry matter) by the plants and hence are termed as macronutrients. These are C H, O, N, P, K, Ca,Mg, and S. Earlier these were also known as major nutrients. Out of these N,P, and K are the primary nutrients and Ca, Mg and S the secondary nutrients and the rest are micronutrient elements.

☆ This classification is based on their relative abundance and prevalence of deficiency and not on their relative importance in plant nutrients.

☆ The deficiency symptoms are generally characteristics and appear on a specific part of plants *i.e.* leaves, stem and roots depending upon its mobility and role in the plants. The deficiency symptoms are used for assessing the fertility of soil and for correcting the nutrient deficiency in question directly in the field without performing costly soil or plant analysis.

☆ Hence the specific and general functions and visual deficiency symptoms of the essential plant nutrients are described to help better understanding the fertilizer or nutritional needs of plant vis-à-vis soil fertility status.

Carbon, Oxygen and Hydrogen

☆ Plants take up C, H, and O (about 90-95 per cent of the plant composition) mainly from the air and water. The C and O are taken up from the atmosphere and possibly as HCO_3 from the soil solution.

☆ Further oxygen is also taken up partly by plants in molecular form (O_2). Hydrogen is taken up as water from the soil solution and also from atmosphere and photolysis.

Function

C,H and O are constituents of organic materials are involved in enzymatic processes and oxidation –reduction process.

Deficiency Symptoms

The deficiency of C,H and O in plants is rare because these nutrients occur in abundance in the nature. The exceptions are extreme conditions of water deficit and water logging under which the uptake of H and O is severely restricted in wilting due to water deficit and yellowing of lower leaves of plants because of O deficiency, which in turn also cause N deficiency and ultimately the death of plants.

Nitrogen

☆ Plants absorb N from the soil solution as nitrate (NO_3) and as ammonium (NH_4) *e.g.* rice.

☆ The content of N in healthy plants ranges between 1 and 5 per cent depending upon the species or variety. Nearly 62 per cent soils are deficient in N.

Function

☆ Nitrogen plays an important role in plant metabolism. As a constituent of chlorophyll it harnesses solar energy and fixes atmosphere CO_2 as carbohydrates. The plant gets stunted and develops chlorosis.

☆ Excess of N produces leathery dark green succulent leaves or crop which delays maturity and increases susceptibility to disease and lodging.

☆ Generally, application of 120-180 kg N/ha to irrigated wheat, rice, maize and 240 N kg/ha to sugarcane and rapeseed-mustard in various soil.

Deficiency Symptoms

☆ Plants containing less than 1.0 per cent N are generally deficient in N. Symptoms of N deficiency first appear on the older leaves.

☆ The N deficient plants are generally stunted develops late and produces less flowers as well as shriveled grains.

Phosphorus

- ✰ The content of P in normal plants ranges from 0.1 per cent to 0.4 per cent. Soil is the main source of P for plant nutrients. Plant root absorb largely as dihydrogen orthophosphate $(H_2PO_4^-)$ ions and under neutral to alkaline condition as orthophosphate $(H_2PO_4^-)$ ions from the soil solution.

- ✰ Further, the plant also absorbs very small amounts of solution organic phosphates *viz.* nucleic acids and phytin. It should preferably be at applied sowing.

Functions

- ✰ Phosphorus plays an important role in energy transformations and metabolic processes in plants. It stimulates plant growth.

- ✰ It is necessary for cell division, merismatic growth root, and seed and fruit development as well as in stimulating flowering, ear emergence and maturation of crop.

Deficiency Symptoms

- ✰ Crop grown on acid soils, calcareous soils, coarse textured soil low in organic matter suffer from P deficiency. In P deficient soil soils the plants develop visual deficiency symptoms, *viz.* they fail to make a quick start, develop poor root system, and remain stunted; being mobile nutrient in plant.

- ✰ P need of crops to produce optimum yield on soil generally 50-60 kg P_2O_5/ha for wheat, 30-40 kg P_2O_5/ha for each crop in rice–wheat system, 20-40 kg P_2O_5/ha for sunflower, safflower and linseed.

Potassium

- ✰ Potassium absorbs K as K^+ ion in large amounts from the soil than any other nutrient (except N and Ca).

- ✰ Plants can store K in large quantities than what is needed for optimum growth without causing toxicity which is termed as luxury consumption. Potassium is deficient in 20 per cent soils of India.

Functions

- ✰ Potassium is involved in regulating the opening and closing of the stomata,(which is the essential for photosynthesis) water and nutrients transport and cooling of plants.

- ✰ It activates a number of enzymes. K helps in maintaining cytoplasmic pH to 7 and, ideal for many enzyme reactions; it activates nearly 60 enzymes. It imparts resistance to plants against fungal and bacterial diseases.

Deficiency Symptoms

✰ Potassium is highly mobile in plants and its deficiency causes on visual symptoms on older leaves. The symptoms of chlorosis start from the leaf margins, followed by scorching and browning of tips and margin in potato, wheat, oat and maize.

✰ Generally 30-40 kg K_2O/ha as basal dressing for irrigated rice wheat, maize *etc.* alleviate its deficiency.

Calcium

Plant roots absorb calcium as Ca^+ ions and its content in healthy plants ranges from 0.2 to 1.0 per cent.

Functions

✰ It is a constituent of the cell wall, an activator of different plants enzymes and is essential for the stability of cell membrane. As calcium is immobile in the plant hence, typical visual deficiency symptoms appear on the younger leaves. Encourage seed production.

✰ Improves intake of other plant nutrients like N and trace elements such as Fe,B,Zn,Cu, and Mn by correcting soil pH

Deficiency Symptoms

✰ Its deficiency is not common. But in acid soil because of its leaching losses and in alkali or sodic soils due to excess Na, the deficiency constraints of Ca are encountered.

✰ Typical visual deficiency symptoms appear on the younger leaves. The Ca-deficiency syndromes in apples are bitter pit and in tomatoes blossom rot, end rots. Lime application to acid soils and gypsum to alkali or sodic soils correct Ca- deficiency.

Magnesium

Plants roots absorb magnesium as Mg^{2+} ion and its content in Mg adequate plants ranges from 0.1 to 0.4 per cent below which the deficiency may occur.

Functions

✰ It is an essential constituent of the chlorophyll as Mg-porphytin. It regulates the activity of several enzyme systems involved in synthesis of nucleic acids and metabolism of carbohydrates.

✰ A large part of magnesium is associated with organic ions *viz.* malate and it also help sin the movement of sugars and translocation of P in plants.

Deficiency Symptoms

✰ Magnesium deficiency is also observed in plants growing in acid soils, leached soils and sandy soils. As Mg is a mobile nutrient in plants, its deficiency symptoms Manifest first on the older leaves and advances upwards to younger leaves as interveinal chlorosis followed by development of purple lesions within the chlorotic and the tissues.

☆ As deficiency proceeds, the veins also become chlorotic and the leaves develop almost uniform pale colour and the purple lesion turn brown orange or red: ultimately the tissues dry up and may die.

Sulphur

☆ Plants absorb Sulphur largely as SO_4^{2-} ion and as SO_2 by the foliage.

☆ The concentration of S in healthy plants ranges from 0.1 to 0.4 per cent almost equal to that of P and Mg.

Functions

☆ It is an essential constituent of the Sulphur-containing amino acids, *viz.* cysteine, cystine and methionine.

☆ It also a constituent of ferrodoxin-containing nitrogenase enzyme, which is involved in biological N fixation (BNF) and electron transfer reaction and in the metabolic activities of vitamins, biotin, thiamine, coenzyme.

☆ A synthesis of glucosides in mustard oil, and in increasing the oil quality of oilseeds crops.

Deficiency Symptoms

☆ Generally crops grown on coarse textured soils under intensive cultivation suffer from its deficiency, especially fertilized with high analysis fertilizer free from S and/or irrigated with tube well and canal water low in S or in rainfed areas.

☆ Soil low in organic matter and coarse textured or leached acid soil are deficient in S. As sulphur is immobile in the plants its deficiency first shows up on the younger leaves as uniform pale green followed by their chlorosis. Depending upon the crop and variety plants having less than 0.1 to 0.2 per cent S contents suffer from its deficiency.

☆ Crop plants having more than 16:1 N:S ratio can be suspected to be deficient in sulphur. Application of 20-60kg S/ha corrected its deficiency in most of these soils/crops.

Micronutrients

☆ These are required in small or micro amount (<50mg/kg dry matter except Fe and Mn) and hence are termed micronutrients.

☆ These are as essential and important like macronutrients and include Fe, Mn, Cu, B, Mo, and Ni. These are subdivided into micronutrient cations (Fe, Mn, Zn, Cu, Ni) and micronutrients anions (B, Mo, and Cl).

Iron

Iron exhibit Fe (II) and Fe(III) oxidation states in plants, and is taken up in the form of Fe^{2+} ion. Its concentration in normal plants ranges between 100 and 50 mg/kg.

Functions

☆ It is a constituent of heme and non heme proteins. These are involved in redox reactions in respiration and photosynthesis.

☆ It is necessary for the synthesis and maintenance of chlorophyll in plants and plays an important role in nucleic acid synthesis. Iron activates a lot of enzymes.

Deficiency Symptoms

☆ Iron deficiency is observed in plants growing on calcareous high pH soils. Iron is immobile in plants usually 30mg Fe^{2+}/kg dry matter is considered the critical level for Fe deficiency.

☆ The iron deficiency shows first on younger leaves as pale yellow interveinal chlorosis of the tissues.

☆ Generally the plants containing less than 50 mg Fe2+/kg dry matter are categorized deficient in iron.

☆ In coarse textured soils organic manures including green manures along with foliar sprays of Fe are more effective in correcting Fe deficiency.

Manganese

☆ Plants absorb Mn as Mn^{2+}, but it is easily oxidized to Mn^{3+} and Mn^{4+} form.

☆ The common Mn carriers that can supply Mn are inorganic salts such as manganese sulphate which is recognized as fertilizer in the FCO.

Functions

☆ It is a constituent of water splitting enzyme associated with O_2 evolution. Manganese is an integral component of the water splitting enzyme associated with photo system II.

☆ It is a constituent of super oxide dismutase (Mn-SOD).Manganese has a role in tricarboxylic acid.

Deficiency Symptoms

☆ Manganese-deficient plants contain less than 25 ppm Mn. Deficiency symptoms of Mn are more severe on middle leaves than on the younger one.

☆ Deficiency symptoms are also observed on plants grown in calcareous soil. In the monocotyledonous plants like cereals deficiency symptoms of Mn appear as greenish grey spot.

☆ The syndromes of Mn deficiency are popularly known as : grey speck of oats, speckled yellow of sugarbeet, marsh spot of peas, *pahala* blight of sugarcane.

Zinc

☆ Plants absorb Zn as zinc ions (Zn^{2+}).

☆ Zinc sufficient plants contain 27 to 150ppm Zn in mature tissues. Since it does not have variable valency and occurs only as Zn(II) oxidation states.

Function

☆ Zinc is a constituent of three enzymes like carbonic anhydrase (CA),alcoholic dehydrogenase (AD) and super oxide dismutase(SOD).

☆ It is involved in the synthesis of indole acetic acid and proteins.

☆ Zn plays an important role in the stabilization and structural orientation of the membrane proteins.

☆ Zn influences translocation and transport of P in plants.

Deficiency Symptoms

☆ Plants containing less than 15 ppm Zn are regarded deficient in plants. Common deficiency symptoms of Zn are interveinal chlorosis, first appearing on the young leaves, and reduction in the size of leaves.

☆ This deficiency is observed in crop growing in soils which are coarse in texture high in pH (sodic),high in $CaCO_3$(calcareous) or low in organic matter or water logging/submerged rice soils. A white to pale yellow chlorosis develop between the mid vein and margin and extent towards the tip, popularly known as white bud in maize or sorghum. The white bud is not a bud but rather a bend or patch of white tissue.

☆ Name given to Zn deficiency symptoms include: Khaira disease of rice reported from India (also called Hadda disease Pakistan, Akagare type II in Japan, Taya-taya and Apaya Popular in Philippines), white bud of maize, mottle leaf (little leaf) or frenching of citrus and little leaf of cotton.

Copper

☆ It exits in plant as cuprous Cu (I) and cupric (II) oxidation states and is taken by plants as Cu^{2+}.

☆ Copper is a transition element exiting in the plants as a component of a large number of proteins and enzymes as Cu (II) and can exchange its valency as per equilibrium:

$$Cu^{2+} + e^- \rightleftarrows Cu^+$$

Function

* Copper is a constituent of number of enzymes or proteins that perform important biochemical function. Copper is important in imparting disease resistance to the plants.
* It enhances the fertility of male flowers.

Deficiency Symptoms

* Plants having less than 5 ppm Cu are regarded as Cu-deficient. Male flowers sterility, delayed flowering and senescence are the most important effects of Cu-deficiency.
* Cu-deficiency is generally observed in crops growing on soils of either inherently low in total Cu or soils high in organic matter (peat soils).
* The plants contain concentration ranging between 5 and 20 mg/kg. The copper is immobile in plants, its deficiency symptoms appear on the young mature leaf and turn pale green with interveinal pale yellow chlorosis at the base of the blade.

Molybdenum

* Mo is the only heavy transition metal taken by the plants as molybdate ions $(MoO2^-_4)$.
* In the plant system it exists as Mo(VI) under oxidative conditions and undergoes reduction to Mo as Mo(V) and Mo(IV)forms.

Functions

* Mo is an essential constituent of many enzymes: nitrogenase that catalyzes biological nitrogen fixation (BNF) by root nodules bacteria of legumes plants.
* Nitrate reductase(NR) reduces nitrate to nitrite and xanthine oxidase/ dehydrogenase which catalyzes the catabolic pathway of purines to uric acid. Mo affects the formation and viability of pollens.

Deficiency Symptoms

* Mo deficiency is observed in plants growing in acid soils. Mo deficiency plants usually contain less than 0.2 mg Mo/kg and the sufficient ones 0.2 to 2 mg mo/kg.
* Its deficiency symptoms appear on old or middle leaves all over surface. Mo deficiency in cauliflower and broccoli is termed as whip-tail.

Boron

* Boron is absorbed by the plants mainly as boric acid (H3BO3) and high pH conditions also as H_2Bo_3.

☆ Most important property of boron is to form stable complexes with organic compounds.

Functions

☆ It is responsible for the cell wall formation and stabilization lignifications and xylem differentiation.

☆ It imparts drought tolerance to the crop. B plays a role in pollen germination and pollen tube growth.

☆ It facilitates ion uptake by way of increasing the activities of plasma – membrane bound H^+-ATPase.

☆ It facilitates trans port of K in guard cells as well as stomal opening.

Deficiency Symptoms

☆ Boron is immobile in plant, and hence its deficiency symptoms develop on new tissue or young leaves as white or transparent lesion in the interveinal mid areas of the leaf.

☆ B deficiency in crops is more critical in highly calcareous, sandy, leached soils, limed acid soils. Its critical limits range between 20 and 70 mg/kg in dicotyledonous and 5-10 mg/kg in graminaceous plants.

☆ The deficiency syndromes are popularly known as brown heart or heart rot in sugar beet and turnip, browning of curds and hollow stem in cauliflower and external and internal cork in apples.

Nickel

☆ Nickel is absorbed by plants as nickel ions (Ni^{2+}).

☆ Normal plants contain 0.1 to 10 mg Ni/kg.

Functions

☆ Nickel is associated with nitrogen metabolism.

☆ It is needed for grain filling and seed vitality.

☆ In free living Rhizobia, adequate Ni supply ensures optimum hydrogenase activity.

☆ It facilitates transport of nutrients to the seed or grain.

Deficiency Symptoms

☆ Critical level of Ni deficiency in barley shoots is 0.1 ppm.

☆ Its deficiency causes accumulation of nitrates and decrease in amino acid content in barley, containing less than 0.1 mg Ni/kg.

Chlorine

- ✰ Chlorine is absorbed as chloride ions (Cl⁻) by the plants.
- ✰ Normal plants have ranging from 100-500 ppm (mg Cl/kg), because of its abundance in nature, its deficiency is noticed only in areas far from the sea where atmosphere deposition does not supply enough Cl and its deficiency have not been reported from any where in India.

Functions

- ✰ It plays a major role in osmoregulation and charge compensation in higher plants. It acts as a cofactor in Mg containing water splitting enzyme of photo system II.
- ✰ Cl in abundance suppresses the plant disease. Chlorine supply improves the nutritional quality of vegetables by preferential lowering the NO_3 –N concentration in tissue.

Deficiency Symptoms

- ✰ Plants having less than 100 ppm Cl is usually designated as deficient.
- ✰ The symptoms appear as irregular yellow to necrotic spotting of leaves, premature wilting, and chlorosis of newly emerging leaves and reduced shoot and root growth.

Identification of Deficiency

- ✰ A brief key to deficiency symptoms developed by Finck (1992) is given below.
- ✰ It is also desirable to understand the role of such factors that might produce symptoms similar to other of the nutrients.

Nutrients	Deficiency Symptoms
Symptoms appear first on older leaves	
Nitrogen	Chlorosis starting from leaf tips
Phosphorus	Reddish colour on green leaves or stem
Potassium	Necrosis on leaf margins
Magnesium	Chlorosis mainly between veins
Manganese	Brownish grayish whitish spot
Symptoms appear first on younger leaves	
Sulphur	Mottled yellow green leaves with yellowish veins
Iron	Mottled yellow green leaves with green veins
Manganese	Brownish black spot
Copper	Youngest leaf has white tip
Boron	Youngest leaf is brownish or dead

Beneficial Elements

☆ **Sodium:** sodium accumulates preferential in the vacuoles and plays a role in maintaining the solute potential of the cell. Sodium can replace some of the essential functions of K in plants. It improves the water balance of plants under limited water supply or under arid climate.

☆ **Silicon:** Silicon has beneficial role in rice and sugarcane crops. It contributes to rigidity and strengthening of the cell wall. It enhances the physiological availability of zinc in plants and counteracts zinc- deficiency –induced phosphorus toxicity.

☆ **Cobalt:** Cobalt enhances N_2 fixation in legumes and improves the nutritional quality of the forage crops for ruminants.

☆ **Selenium:** The chemistry of Se is similar to that of S with Se existing in Se^{2+}, Se^{4+} and Se^{6+} states. Se is an essential element for animals. To avoid Se- deficiency in animals the concentration of Se in diets should be 0.1 to 0.3 mg/kg dry matter. In livestock symptoms of Se- deficiency include white muscle disease or nutritional muscular dystrophy.

☆ **Aluminium:** In plants species having high tolerance and capacity for Al, the beneficial effects of Al on growth accrue from its ability to alleviate toxicity of P, Zn and Cu.

☆ **Vanadium:** Vanadium (V) favours nitrogen fixation in leguminous plants. The chemical behavior of V and Mo is similar. It is brought that V might partly substitute the function of Mo in nitrogenase system of *Azotobacter.* Other elements being investigated for the beneficial effects on plants including Bromine, iodine, lead, cadmium, chromium, fluorine *etc.* but no evidence has been found so far.

Nutrient Movement in Soil

☆ Transpiration of water by plants induces the movement of water from soil to plants. Nutrients which are dissolved in soil solution move to the plant roots with water in response to the hydraulic gradient by created by the roots as a result of continuous absorption of water. However, differential requirements of nutrients vis-à-vis water by plants result in either accumulation or depletion of nutrients in the vicinity of root surface.

☆ There are two separate processes involved in the movement of nutrients to plant roots. Movement of nutrients with water is known as 'mass flow' and the movement through water as 'diffusion'. In addition, growing plants roots move into the soil and displace it. In the process, roots come in contact with the soil and intercept the nutrients from the soils through contact exchange. Thus, three mechanisms recognized for nutrient movement are–mass flow, diffusion and root inception.

Mass Flow

☆ Mass flow is the movement of nutrients through the soil to the root in the convective flow of water caused by the plant water absorption.

☆ The nutrients move physically with the flow of soil water towards plant root, because continuous absorption of water due to transpiration creates a hydraulic gradient at the root surface, where their absorption through the roots takes place along with water.

☆ Some nutrients ions that move freely with water, generally for large distances, are called mobile nutrients However, others move slowly with the water for relatively small distance, only a few millimeters, called immobile nutrients.

☆ The nutrient uptake through this mechanism is directly related to transpiration-the amount of water used by plants. Mass flow is responsible for supplying the root with much of the plant needs for N, S, Cl, Ca, and Mg when presents in sufficient concentration in the soil solution, but not for relatively immobile nutrient elements *e.g.* P, K, Zn *etc.*

Diffusion

☆ When mass flow is unable to supply sufficient quantities of a nutrient and a continued uptake occurs, the concentration of the nutrients at the root surface is reduced and a concentration gradient is established.

☆ Ions move by Brownian movement from points of higher concentration to those of lower concentration when a concentration gradient exists. Movement of a nutrient ion in response to the concentration gradient is termed as diffusion.

Root Interception

☆ The term root interception coined by Stanley A. barber of Purdue State University, USA,is used to describe the soil nutrients at the root surface that do not have to move to the interface to be positionally available for absorption but are approached by the roots itself in the soil.

☆ The quantity of the nutrients supplied by root-interception is taken as the quantity present in a volume of soil equal to the root volume. It is generalized that root volume for several crop species growing in soil is less than one percent of the soil volume in 0-20 cm depth. Hence, calculations are based on the volume.

☆ Conceptually, as a root system develops and exploits more soil, soil solution and soil surface retaining absorbed ions are exposed to the root mass and adsorption of these ions occurs by a contact exchange mechanism of Jenny and Overstreat (1939). Root interception of the nutrients in fertile soils can be enhanced by the mycorrhiza, a symbiotic association between certain fungi and plant roots.

Chapter 7

Irrigation Management

Introduction

⭐ Irrigation is the process of applying water to the crops artificially to fulfill their water requirements. Nutrients may also be provided to the crops through irrigation. The various sources of water for irrigation are wells, ponds, lakes, canals, tube-wells and even dams. Irrigation offers moisture required for growth and development, germination and other related functions.

⭐ The frequency, rate, amount and time of irrigation are different for different crops and also vary according to the types of soil and seasons. For example, summer crops require a higher amount of water as compared to winter crops.

Water Resources

⭐ For an estimated explanation of where Earth's water exists, look at the chart below. By now, you know that the water cycle describes the movement of Earth's water, so realize that the chart and table below represent the presence of Earth's water at a single point in time

⭐ The distribution of water on, in, and above the Earth

⭐ The water resources can be divided into two equal parts: Surface Water Resources and Sub Surface Water Resources.

❖ Surface water is water on the surface of the planet such as in a river, lake, stream, reservoirs, wetland, or ocean. It can be contrasted with groundwater and atmospheric water. The main uses of surface water include drinking-water and other public uses, irrigation uses, and for use by the thermoelectric-power industry to cool electricity-generating equipment.

☆ Sub-surface water resources include groundwater. Groundwater is an important part of the water cycle.

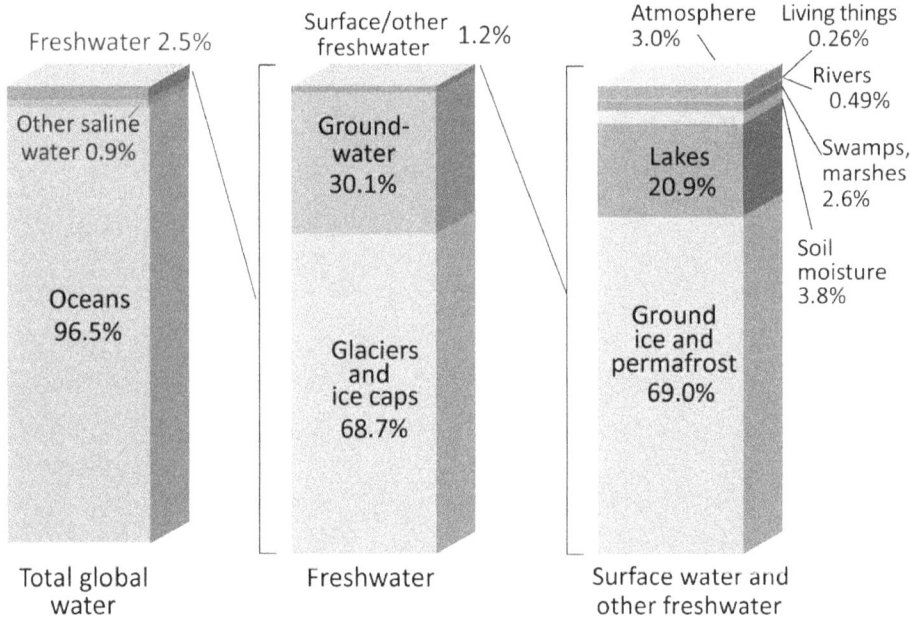

Credit: U.S. Geological Survey, Water Science School. https://www.usgs.gov/special-topic/water-science-school
Data source: Igor Shiklomanov's chapter "World fresh water resources" in Peter H. Gleick (editor), 1993, Water in Crisis: A Guide to the World's Fresh Water Resources. (Numbers are rounded).

Soil-Plant-Water Relationship

☆ The plant needs soil and water for its growth. The soil anchors the plant and allows the root system to grow. Soil consists of a three-phase system – solid, liquid and gas.

☆ The complex mass of organic and mineral matter acts as a matrix consisting of a number of pores of various sizes.

☆ The void space within the particles is known as soil pore space. The pores of the soil hold moisture attached to the soil particle by surface tension. If water is present in amounts more than that can be stored, the soil expels air.

☆ On the other hand, in dry soils, the water molecules are present as a thin film around the soil particles.

Classification of Soil-water

Water occurs in varying proportions in the soil pores. Based on the physical factors, the water held in the pores can be defined as:

☆ When a soil sample is saturated with water and allowed to drain the amounts that are present in excess, the gravitational water is drained off easily and rapidly from the root zone. Gravitational water is not available for use by plants

☆ The amount of water retained in the soil by surface tension after the draining of gravitational water is known as capillary water. Capillary water is the main source of water for plant growth

☆ Hygroscopic water is the moisture content absorbed and retained by dry soil as a thin film over the soil particles. Neither this water is available to plants.

Role of Water in Plant Growth

Water plays a key role in plant growth. The most important functions of water are listed below:

☆ Water constitutes 85-90 per cent of the body weight of younger plants and 20-50 per cent of the weight of mature plants

☆ Oxygen and hydrogen required for carbohydrate synthesis during the photosynthesis is provided by water

☆ Nutrients required for the plant growth are transported from soil to the plant with the help of water

☆ Food synthesized in the leaves gets dissolved in the water and then distributed throughout the plant body

☆ Water is essential for optimal transpiration. Transpiration helps maintain the plant temperature through dissipation of heat.

Soil Plant Atmosphere Continuum

☆ The pathway through which water moves from the soil to the atmosphere via plant is known as the soil-plant-atmosphere continuum. The atmosphere has low water potential while a relatively higher potential occurs inside the leaves. This creates a diffusion gradient across the pores in stomata of the leaves leading to vaporization of water molecules present in the leaves.

☆ As water vapour transpires out of the leaves, it is replaced with water molecules maintained inside the leaves at saturation vapour pressure. Xylem cells of plant pull water from the roots towards the leaf.

Hydrological Cycle

The hydrological cycle of the earth is the sum total of all processes in which water moves from the land and ocean surface to the atmosphere and back in form of precipitation. The hydrological cycle is dependent on various factors and is equally affected by oceans and land surfaces.

The hydrologic cycle consists of 4 key components:

☆ Precipitation

☆ Runoff

☆ Storage

☆ Evapotranspiration

Earth's water is always in movement, and the natural water cycle, also known as the hydrologic cycle, describes the continuous movement of water on, above, and below the surface of the Earth. Water is always changing states between liquid, vapor, and ice, with these processes happening in the blink of an eye and over millions of years.

Precipitation

☆ Precipitation occurs when atmospheric moisture becomes too great to remain suspended in clouds.

☆ It denotes all forms of water that reach the earth from the atmosphere, the usual forms being rainfall, snowfall, hail, frost and dew.

☆ Once it reaches the earth's surface, precipitation can become surface water runoff, surface water storage, glacial ice, water for plants, groundwater, or may evaporate and return immediately to the atmosphere. Ocean evaporation is the greatest source (about 90 per cent) of precipitation.

Runoff

☆ Runoff is the water that flows across the land surface after a storm event.

☆ As the flow bears down, it notches out rills and gullies which combine to form channels. These combine further to form streams and rivers.

☆ The geographical area which contributes to the flow of a river is called a river or a watershed.

Storage

☆ Portion of the precipitation falling on land surface which does not flow out as runoff gets stored as either as surface water bodies like Lakes, Reservoirs and Wetlands or as subsurface water body, usually called Ground water.

☆ Ground water storage is the water infiltrating through the soil cover of a land surface and traveling further to reach the huge body of water underground.

☆ The amount of ground water storage is much greater than that of lakes and rivers.

Evapotranspiration

☆ Evapotranspiration is actually the combination of two terms – evaporation and transpiration.

☆ The first of these, that is, evaporation is the process of liquid converting into vapour, through wind action and solar radiation and returning to the atmosphere.

☆ Evaporation is the cause of loss of water from open bodies of water, such as lakes, rivers, the oceans and the land surface.

☆ Transpiration is the process by which water molecules leaves the body of a living plant and escapes to the atmosphere.

☆ Evapotranspiration, therefore, includes all evaporation from water and land surfaces, as well as transpiration from plant

Irrigation

Irrigation is defined as the artificial application of water to the soil for the purpose of crop growth or crop production in supplement to rainfall and ground water contribution.

Irrigation Management

☆ Irrigation water management is the act of timing and regulating irrigation water applications in a way that will satisfy the water requirement of the crop without the waste of water, soil, plant nutrients, or energy.

☆ It means applying water according to crop needs in amounts that can be held in the soil available to crops and at rates consistent with the intake characteristics of the soil and the erosion hazard of the site.

Seasons of Rainfall in India

Winter (Cold dry period)	January—February
Summer (Hot weather period)	March — May
Kharif (South-West monsoon) *70% of the rainfall	June — September
Rabi (North-East monsoon) *Tamil Nadu receives its 60 per cent of rainfall from NEM	October — December

Water Budget

Water budget can be defined as the relationship between the inflow and outflow of water through a specified region or a country. It gives a comparison between the supply and demand of water, making it possible to identify periods of excess and deficit precipitation.

India's Water Budget

Total geographical area	328M.ha.	
Average annual rainfall	1190mm	
In million hectare metre	1190 x 328	390 M ha m
Contribution from snowfall		8M ha m
TOTAL		**398 (400 M ha m approx.)**

☆ The rainfall below 2.5 mm is not considered for water budgeting, since it will immediately evaporate from surface soil without any contribution to surface water or ground water.

☆ There are on an average 130 rainy days in a year in the country out of which the rain during 75 days considered as effective rain. The remaining 55 days are very light and shallow which evaporates immediately without any contribution to surface or ground water recharge.

☆ Considering all these factors it is estimated that out of 400 million hectare meter of annual rainfall 70 million hectare meter is lost to atmosphere through evaporation and transpiration, about 115 million hectare meter flows as surface run-off and remaining 215 million hectare meter soaks or infiltrates into the soil profile.

Total surface run-off has been estimated by Irrigation Commission of India in 1972 as follows:

Rain fall contribution	115
Contribution from outside the country through streams and rivers	20
Contribution from regeneration from ground water in Stream and rivers	45
Total Surface Run-off	**180**

Irrigation Methods Suitable for Different Crops

Irrigation Method	*Crops*
Flooding	Rice and jute
Check basin	Groundnut, pulses, finger millet
Border strip	Close growing crops tobacco, potato, sorghum, sugarcane, vegetables

Irrigation Method	Crops
Furrow	Cotton, maize,
Surge	Maize, sorghum
Corrugation (shallow and small furrow)	Wheat, groundnut, setaria sp.
Drip	Sugarcane
Sprinkler	Vegetable and fruit crop

Crop Water Requirement

✫ Water requirements of a crop is the quantity of water needed for normal crop growth and yield in a period of time to a place and may be supplied by precipitation or by irrigation or by both.

✫ Water is needed mainly to meet the demand of evaporation (E), transpiration (T) and metabolic activity of plant together known as Consumptive Use (C.U).

✫ So, water requirement = IW + ER + S

IW - Irrigation Water, in cm;

ER – Effective Rainfall, in cm;

S – Soil profile contribution.

Water Requirement of Crops

Rice

✫ The daily consumptive use of rice varies from 6-10 mm and total water is ranges from 1100 to 1250 mm depending upon the agro climatic situation.

✫ Of the total water required for the crop, 3 per cent or 40 mm is used for the nursery, 16 per cent or 200 mm for the land preparation *i.e.* puddling and 81 per cent or 1000 mm for field irrigation of the crop.

✫ The growth of rice plant in relation to water management can be divided into four periods *viz.*, Seedling, vegetative, reproductive and ripening.

Groundnut

✫ Evapotranspiration is low during the first 35 days after sowing and last 35 days before harvest and reaches a peak requirement between peg penetration and pod development stages. Total water requirement 500-550 mm.

✫ After the sowing irrigation the second irrigation can be scheduled 25 days after sowing *i.e.* 4 or 6 days after first hand hoeing and thereafter irrigation interval of 15 days is maintained upto peak flowering.

☆ During the critical stages the interval may be 7 or 10 days depending upon the soil and climate. During maturity period the interval is 15 days.

Finger Millet/Ragi

☆ Finger millet is a drought tolerant crop. Pre-planting irrigation at 7 or 8 cm is given. Total water requirement: 350 mm

☆ Third day after transplantation life irrigation with small quantity of water is sufficient for uniform establishment.

☆ Water is then withheld for 10-15 days after the establishment of seedling for healthy and vigorous growth. Subsequently three irrigations are essential at primordial initiation, flowering and grain filling stages.

Sugarcane

☆ Total water requirement: 1800-2200 mm. Formative phase (120 days from planting) is the critical period for water demand.

☆ To ensure uniform emergence and optimum number of tillers per unit area lesser quantity of water at more frequencies is preferable.

☆ The response for applied water is more during this critical phase during which the crop needs higher quantity of water comparing, the other two phases. Water requirement, number of irrigations *etc.*, are higher during this period.

Maize

☆ Total water requirement: 500 – 600 mm. The water requirement of maize is higher, but it is very efficient in water use.

☆ Growth stages of maize crop are sowing, four leaf stage, knee high, grand growth, tasseling, silking early dough and late dough stages.

☆ Crop uniformly requires water in all these stages. Of this, tasseling, silking and early dough stages are critical periods.

Cotton

☆ Total water requirement: 550 – 600 m. Cotton is sensitive to soil moisture conditions.

☆ Little water is used by plant with early part of the season and more water is lost through evaporation than transpiration.

☆ As the plant grows, the use of water increases from 3 mm/day reaching a peak of 10 mm a day when the plant is loaded with flowers and boll.

☆ Water used during the emergence and early plant growth is only 10 per cent of the total requirement. Ample moisture during flowering and boll development stages is essential.

☆ In the early stage as well as at the end the crop requires less water. Water requirement remains high till the boll development stage.

Sorghum

☆ Total water requirement: 350-500 mm

Pulses

☆ Total water requirement: 200-450 mm

☆ Mostly the pulses are grown under rainfed condition.

☆ Some pulse crops like Redgram, Blackgram, Greengram are grown in summer season as irrigated crop which need 3 to 4 irrigation at critical stags like germination, flowering and pod formation.

Critical Stages of Crops for Irrigation

Sl.No.	Cereals	Critical Stages of Crops for Irrigation
1	Rice	Tillering, panicle, initiation, heading and flowering
2	Wheat	CRI, Tillering, Late joining, flowering, milking and dough
3	Maize	Tasseling and silking to dough stage
4	Sorghum	Booting, blooming, milking and dough stage
5	Pearl millet	Heading and flowering
6	Finger millet	Primordial initiation and flowering
	Pulses	
1	Chickpea	Late vegetative phase and pod development
2	Pea	Flowering and early pod formation
3	Blackgram	Flowering and pod setting
4	Greengram	Flowering and pod setting
5	Lucern	After cutting and flowering
6	Beans	Flowering and pod settings
	Oilseed	
1	Groundnut	Flowering, peg formation and pod development
2	Soybean	Blooming and seed formation
3	Sunflower	Buttoning, knee high, flowering and early seed formation
4	Sesamum	Blooming to maturity

Water Use Efficiency (WUE)

It is the yield of a marketable crop produced per unit of water used in evapo-transpiration or it is the dry matter produced per unit of water used and it is expressed as kg/ha^{-mm} (cm).

Water Use Efficiency is of two types:

* **Field Water Use Efficiency:** Field WUE = Crop Yield (kg/ha)/(ET + S + D)

  Where ET: Evapotranspiration loss of water;

  S: Ground water contribution;

  D: Deep Percolation losses;

* **Crop Water Use Efficiency:** Crop WUE = Crop Yield (kg/ha)/(E + T +G)

  E: Evaporation loss;

  T: Transpiration loss;

  G: Metabolic use of plant;

Water Use Efficiency of Major Field Crops

Sl.No.	Crop	WUE (kg/ha mm)
1	Finger Millet	**13.4**
2	Wheat	12.6
3	Groundnut	9.2
4	Sorghum	9.0
5	Pearl millet, maize	8.0
6	Rice	3.7 (lowest)

Irrigation Efficiency

It is defined as the ratio of water output to the water input, *i.e.,* the ratio or percentage of the irrigation water consumed by the crop of an irrigated farm, field or project to the water delivered from the source.

$$Ei = \frac{Wc}{Wr} \times 100$$

where,

Ei = irrigation efficiency (per cent)

Wc = irrigation water consumed by crop during its growth period in an irrigation project.

Wr = water delivered from canals during the growth period of crops.

In most irrigation projects, the irrigation efficiency ranges between 12 to 34 per cent.

Irrigation Water Quality

* Irrigation water quality refers to the kind and amount of salts present in the water and their effects on crop growth and development.

☆ High salt concentrations influence osmotic pressure of the soil solution and affect the ability of plants to absorb water through their roots.

☆ However, an appropriate evaluation of the water quality prior to its use in irrigation will help in arresting any harmful effect on plant productivity and ground water recharge.

☆ The suitability of water for irrigation is determined in several ways including the degree of acidity or alkalinity (pH), EC (Electrical Conductivity), Residual Sodium Carbonate (RSC), Sodium Adsorption Ratio (SAR), Permeability Index (PI) and Total Hardness (TH) along with the effects of specific ions.

☆ The assessment of water quality criteria for irrigation is based on the consideration of the related aspects like the possible effects on the physico-chemical properties of the soil and the impact on crop yield

Classification of Irrigation Water Quality

Various standards for determining the Irrigation water quality:

Residual Sodium Carbonate

Carbonate associates quickly with Ca and Mg and form $CaCO_3$ and $MgCO_3$. The Na replaces Ca and Mg and synthesizes Na_2CO_3 which again causes sodium hazard (called as Residual Sodium Carbonate RSC).

RSC in Water (m.eq/l)	Suitability for Irrigation	Remarks
>2.5	Not suitable for irrigation	Needs gypsum
1.25-2.5	Marginal	Need gypsum
Less than 1.25	Safe	-

Sodium Hazard of Irrigation Water

SAR is commonly used as an index for evaluating the sodium hazard associated with an irrigation water supply. The SAR is defined as the square root of the ratio of the sodium (Na) to calcium + magnesium (Ca + Mg).

SAR= [Na+]/ √Ca + MgC

where all cation measurements are expressed in millimoles per liter (mmol/L). Alternatively, if the cation measurements are expressed in milliequivalents per liter (meq/L), then the SAR is defined to be:

$$SAR = \left[Na^+ \right] / \frac{\sqrt{\left[Ca^{2+} \right] + \left[Mg^{2+} \right]}}{2}$$

Irrigation waters having high SAR levels can lead to the build up of high soil Na levels over time, which in turn can adversely effect soil infiltration and poor aeration.

Sodium Hazard	Class	SAR
Low	S1	<10
Medium	S2	10-18
High	S3	18-26
Very High	S4	26-31

Boron Hazard of Irrigation Water

Class		Boron (ppm)	Suitability
Normal water	C_1	<3	Ideal for all crops on all soils
Low boron water	C_2	3-4	All crops on heavy and medium soils
Medium boron water	C_3	4-5	Can be used for most crops on heavy soils
Boron water	C_4	5-10	Semi-tolerate and tolerate crops on heavy soils
High boron water	C_5	>10	Not suitable for irrigation

Classification of Irrigation Water Quality

Quality of Water	EC (m. mhos/ cm)	pH	Na (per cent)	Cl (me/l)	SAR
Excellent	0.5	6.5 - 7.5	30	2.5	1.0
Good	0.5 - 1.5	7.5 - 8.0	30 - 60	2.5 - 5.0	1.0 - 2.0
Fair	1.5 - 3.0	8.0 - 8.5	60 - 75	5.0 - 7.5	2.0 - 4.0
Poor	3.0 - 5.0	8.5 - 9.0	75 - 90	7.5 - 10.	4.0 - 8.0
Very poor	5.0 - 6.0	9.0 - 10.	80 - 90	10.0 - 12.5	8.0 - 15.0
Unsuitable	>6.0	> 10	>90	>12.5	>15

Methods of Irrigation

The four methods of irrigation are:

    I. Surface

   II. Sprinkler

  III. Drip/trickle

  IV. Subsurface

I. Surface Irrigation

☆ In all the surface methods, Surface irrigation Uncontrolled flooding, Border strip, Check, Basin, Furrow method. of irrigation, water is either ponded on the soil or allowed to flow continuously over the soil surface for the duration of irrigation. Although surface irrigation, Surface irrigation Uncontrolled flooding, Border strip, Check, Basin, Furrow method.

✰ It is the oldest and most common method of irrigation; it does not result in high levels of performance. This is mainly because of uncertain infiltration rates which are affected by year-to-year changes in the cropping pattern, cultivation practices, climatic factors, and many other factors. As a result, correct estimation of irrigation efficiency of surface irrigation is difficult. Application efficiencies for surface methods may range from about 40 to 80 per cent.

Surface irrigation consists of a broad class of irrigation methods in which water is distributed over the soil surface by gravity flow. The irrigation water is introduced into level or graded furrows or basins, using siphons, gated pipe, or turnout structures, and is allowed to advance across the field. Surface irrigation is best suited to flat land slopes, and medium to fine textured soil types which promote the lateral spread of water down the furrow row or across the basin.

Surface Irrigation.
(*Source*: NRCS http://photogallery.nrcs.usda.gov)

✰ Surface irrigation which includes the following:

❖ Uncontrolled (or wild or free) flooding method,

❖ Border strip method,

❖ Check method,

❖ Basin method, and

❖ Furrow method.

(a) Uncontrolled Flooding

✰ When water is applied to the cropland without any preparation of land and without any levees to guide or restrict the flow of water on the field, the method is called 'uncontrolled', wild or 'free' flooding. In this method of flooding, water is brought to field ditches and then admitted at one end of the field thus letting it flood the entire field without any control.

☆ Uncontrolled flooding generally results in excess irrigation at the inlet region of the field and insufficient irrigation at the outlet end. Application efficiency is reduced because of either deep percolation (in case of longer duration of flooding) or flowing away of water (in case of shorter flooding duration) from the field. The application efficiency would also depend on the depth of flooding, the rate of intake of water into the soil, the size of the stream, and topography of the field.

☆ Obviously, this method is suitable when water is available in large quantities, the land surface is irregular, and the crop being grown is unaffected because of excess water. The advantage of this method is the low initial cost of land preparation. This is offset by the disadvantage of greater loss of water due to deep percolation and surface runoff.

Advantages

This method has the following advantages:

☆ No other field layout become necessary except the levelling and preparation of impervious boundary bunds.

☆ No land area is utilized for water distribution. As a result, wastage of land becomes minimum

☆ Labour supervision is required for application of water in the field.

☆ Labour requirement is minimum.

Disadvantages

Following are disadvantages of check basin method:

☆ It is the most inefficient methods of irrigation as only about 20 per cent of water is actually used by plants and the rest is being lost as runoff, seepage and evaporation

☆ Levelling of land increases cost of cultivation.

☆ This method is unsuitable for crops that are sensitive to water logging.

☆ Crop growth is not uniform as the water distribution by this method is very uneven.

☆ There is a possibility of soil erosion.

(b) Border Strip Method

☆ Border strip irrigation (or simply 'border irrigation') is a controlled surface flooding method of applying irrigation water. In this method, the farm is divided into a number of strips which can be 3-20 metres wide and 100-400 metres long. These strips are separated by low levees (or borders). The strips are level between levees but slope along the length according to natural slope. If possible, the slope should be between 0.2 and 0.4 per cent. But, slopes as flat as 0.1 per cent and as steep as 8 per cent can also be used.

☆ In case of steep slope, care should be taken to prevent erosion of soil. Clay loam and clayey soils require much flatter slopes (around 0.2 per cent) of the border strips because of low infiltration rate. Medium soils may have slopes ranging from 0.2 to 0.4 per cent. Sandy soils can have slopes ranging from 0.25 to 0.6 per cent.

☆ Water from the supply ditch is diverted to these strips along which it flows slowly towards the downstream end and in the process it wets and irrigates the soil. When the water supply is stopped, it recedes from the upstream end to the downstream end.

☆ The border strip method is suited to soils of moderately low to moderately high intake rates and low erodibility. This method is suitable for all types of crops except those which require prolonged flooding which, in this case, is difficult to maintain because of the slope. This method, however, requires preparation of land involving high initial cost.

Advantages

This method has the following advantages:

☆ It is the best method to irrigate close growing crops.

☆ Uniform distribution and high water application efficiencies are possible if the system is properly designed.

☆ Labour requirement is less to irrigate the field.

☆ Operation of this system is simple and easy.

☆ Excess rainwater is drained out if outlets are available.

Disadvantages

Following are disadvantages of check basin method:

☆ More labour is required for leveling of the field.

☆ Ridges cut down the neat cropped area.

☆ Large irrigation streams are required.

☆ Repair of ridges and supervision during irrigation are needed.

(c) Check Method

☆ The check method of irrigation is based on rapid application of irrigation water to a level or nearly level area completely enclosed by dikes. In this method, the entire field is divided into a number of almost levelled plots (compartments or 'Kiaries') surrounded by levees. Water is admitted from the farmer's watercourse to these plots turn by turn.

☆ This method is suitable for a wide range of soils ranging from very permeable to heavy soils. The farmer has very good control over the distribution of water in different areas of his farm.

☆ Loss of water through deep percolation (near the supply ditch) and surface runoff can be minimised and adequate irrigation of the entire farm can be achieved. Thus, application efficiency is higher for this method. However, this method requires constant attendance and work (allowing and closing the supplies to the levelled plots).

☆ Besides, there is some loss of cultivable area which is occupied by the levees. Sometimes, levees are made sufficiently wide so that some 'row' crops can be grown over the levee surface.

Advantages

This method has the following advantages:

☆ It is the best method of irrigation for leveled fields.

☆ It does not require any technical knowledge.

☆ This method is more useful in soils having lesser infiltration.

☆ In this method, rain water stays in basins, hence soil erosion is not caused.

☆ It has lesser economic investment.

☆ It irrigates more area and Crops gets sufficient water.

Disadvantages

Following are disadvantages of check basin method:

☆ Due to seepage in drains, wastage of water is caused.

☆ Machines cannot be used m this method because during spray of insecticides or fertilizers, the earthen walls of basins are damaged.

☆ There IS imbalance in distribution of labour. After growth of crops, water reaches the basins in disproportionate quantity thereby causing wastage of water.

☆ Creation of problem of water logging.

(d) Basin Method

☆ This method is frequently used to irrigate orchards. Generally, one basin is made for one tree. However, where conditions are favourable, two or more trees can be included in one basin.

(e) Furrow Method

☆ In the surface irrigation methods discussed above, the entire land surface is flooded during each irrigation. An alternative to flooding the entire land surface is to construct small channels along the primary direction of the movement of water and letting the water flow through these channels which are termed 'furrows', 'creases' or 'corrugation'.

☆ Furrows are small channels having a continuous and almost uniform slope in the direction of irrigation. Water infiltrates through the wetted perimeter of the furrows and moves vertically and then laterally to saturate the soil. Furrows are used to irrigate crops planted in rows.

☆ Furrow lengths may vary from 10 metres to as much as 500 metres, although, 100 metres to 200 metres are the desirable lengths and more common. Very long furrows may result in excessive deep percolation losses and soil erosion near the upstream end of the field. Preferable slope for furrows ranges between 0.5 and 3.0 per cent. Many different classes of soil have been satisfactorily irrigated with furrow slope ranging from 3 to 6 per cent. In case of steep slopes, care should be taken to control erosion. Spacing of furrows for row crops (such as corn, potatoes, sugarbeet, *etc.*) is decided by the required spacing of the plant rows. The furrow stream should be small enough to prevent the flowing water from coming in direct contact with the plant. Furrows of depth 20 to 30 cm are satisfactory for soils of low permeability. For other soils, furrows may be kept 8 to 12 cm deep.

☆ Water is distributed to furrows from earthen ditches through small openings made in earthen banks. Alternatively, a small-diameter pipe of light weight plastic or rubber can be used to siphon water from the ditch to the furrows without disturbing the banks of the earthen ditch.

☆ Furrows necessitate the wetting of only about half to one-fifth of the field surface. This reduces the evaporation loss considerably. Besides, puddling of heavy soils is also lessened and it is possible to start cultivation soon after irrigation. Furrows provide better on-farm water management capabilities for most of the surface irrigation conditions, and variable and severe topographical conditions. For example, with the change in supply conditions, number of simultaneously supplied furrows can be easily changed. In this manner, very high irrigation efficiency can be achieved.

Advantages

The followings are the major advantages of furrow irrigation.

☆ A quick mass area irrigation is possible.

☆ Time and Labour saving method.

☆ Low investment required to buy equipment.

☆ This is a cost-efficient method as it minimizes water loss of gravity irrigation system.

☆ The unit cost of pumped water is lower which saves money.

☆ Recirculating irrigation runoff water is possible.

✫ It is possible to reduce chemical leaching in furrow irrigation.

✫ Higher crop yields can be ensured through proper furrow irrigation practices.

Disadvantages

The followings are the disadvantages of furrow irrigation.

✫ Possibility of increased salinity between furrows,

✫ Loss of water at the downstream end unless end dikes are used,

✫ The necessity of one extra tillage work, *viz.*, furrow construction,

✫ Possibility of increased erosion, and

✫ Furrow irrigation requires more labour than any other surface irrigation method.

✫ Not Suitable for sandy soil.

II. Sprinkler Irrigation

✫ Sprinkler irrigation is a method of irrigation in which water is sprayed, or sprinkled through the air in rain like drops.

✫ The spray and sprinkling devices can be permanently set in place (solid set), temporarily set and then moved after a given amount of water has been applied (portable set or intermittent mechanical move), or they can be mounted on booms and pipelines that continuously travel across the land surface (wheel roll, linear move, center pivot).

Sprinkler Irrigation.
(*Source*: https://civilseek.com/methods-of-irrigation).

☆ Sprinkler irrigation offers a means of irrigating areas which are so irregular that they prevent use of any surface-irrigation methods.

☆ By using a low supply rate, deep percolation or surface runoff and erosion can be minimized.

Advantages

The major advantages of the sprinkler irrigation system are summarised as under:

☆ It is affordable and completely easy to set up. You will not need to spend much on labour cost for setting it up.

☆ There is no requirement of using many areas of your field for setting up the sprinkler irrigation.

☆ The interference with cultivation for setting up the sprinkler irrigation is very less. So, you will not face a huge loss.

☆ Frequent application of water can be supplied to the plants you will not need to do it yourself. The water distribution will always be equal.

☆ The amount of water being supplied can be controlled so you will be also able to save water depending on the necessity and requirements of plants.

☆ The sprinkler irrigation is suitable for setting up in all types of soil can be used for other purposes as well such as cooling during high temperature.

Disadvantages

The disadvantages of the sprinkler irrigation system described as under:

☆ The cost investment cost required for purchasing the equipment of the sprinkler irrigation system is high.

☆ Using the sprinkler irrigation for supplying saline water can result in problems to arise.

☆ There is a chance of water getting evaporated from the sprinkler irrigation when the surrounding environment is windy and high in terms of humidity.

☆ There is a chance of the nozzles of the sprinklers getting clogged due to the deposit of debris and sediments from water that is used.

☆ For spraying water droplets evenly there is a requirement of constant water supply.

☆ There is a requirement of continuous power supply for operating the sprinkler irrigation system.

III. Drip/Trickle Irrigation Systems

☆ Drip irrigation, also referred to as micro irrigation, trickle irrigation or localized irrigation, involves dripping water onto the soil at very low rates (2 - 20 liters/hour) from a system of small diameter plastic pipes fitted with outlets called emitters or drippers.

☆ Water is applied close to plants so that only part of the soil in which the roots grow is wetted, unlike surface and sprinkler irrigation, which involves wetting the entire soil profile.

☆ Drip/trickle irrigation systems are methods of micro- irrigation wherein water is applied through emitters to the soil surface as drops or small streams.

☆ The discharge rate of the emitters is low so this irrigation method can be used on all soil types.

Drip/Trickle Irrigation Systems.
(*Source*: https://civilseek.com/methods-of-irrigation)

Advantages

Drip irrigation provides a large number of advantages over other irrigation methods:

☆ Extensive land leveling and bunding is not required, drip irrigation can be employed in all landscapes;

☆ Irrigation water can be used at a maximum efficiency level and water losses can be reduced to a minimum;

☆ Soil conditions can be taken into account to a maximum extent and soil erosion risk due to irrigation water impact can be reduced to a minimum;

☆ Fertilizer and nutrients can be used with high efficiency; as water is applied locally and leaching is reduced, fertilizer/nutrient loss is minimized (reduced risk of groundwater contamination);

☆ Weed growth is reduced as water and nutrients are supplied only to the cultivated plant;

☆ Positive impact on seed germination and yield development;

☆ Low operational costs due to reduced labor requirement, in particular energy cost can be reduced as drip irrigation is operated with lower pressure than other irrigation methods.

Disadvantages

The disadvantages of the sprinkler irrigation system described as under:

☆ High initial investment requirements;

☆ Regular capital requirement for replacement of drip irrigation equipment on the surface (damage due to movement of equipment, UV-radiation);

☆ Drip irrigation emitters are vulnerable to clogging and dysfunction (water filters required, regular flushing of pipe system);

☆ High skill requirements for irrigation water management in order to achieve optimal water distribution; Soil salinity hazard.

IV. Subsurface Irrigation

☆ Subsurface drip irrigation systems are highly efficient irrigation systems that apply accurate amounts of water directly to the root zone, preventing water loss through evaporation and other negative effects of surface irrigation.

☆ This is especially suitable for arid, semi-arid, hot, and windy areas with limited water supply. However, as the system is relatively complex and most likely automated, it is more suitable for medium to large-scale production.

☆ Subsurface irrigation consists of methods whereby irrigation water is applied below the soil surface. The specific type of irrigation method varies depending on the depth of the water table.

☆ When the water table is well below the surface, drip or trickle irrigation emission devices can be buried below the soil surface (usually within the plant root zone).

☆ The irrigation water should be of good quality to avoid excessive soil salinity.

☆ Sub irrigation results in a minimum of evaporation loss and surface waste and requires little field preparation and labor.

Subsurface Irrigation
(*Source*: https://civilseek.com/methods-of-irrigation)

Advantages

Subsurface drip irrigation provides a large number of advantages over other irrigation methods:

- ☆ High degree of control over water application with the potential for high uniformity of application
- ☆ Evaporation is reduced
- ☆ The amount of water can be fine-tuned. This avoids water loss caused by run off or evaporation
- ☆ Frequent irrigation allows for optimum soil moisture content in the root zone
- ☆ Great performance in windy and arid locations
- ☆ If pre-treated wastewater is used for irrigation, the risk of direct contact with crops and laborers is reduced

Disadvantages

The disadvantages of the Subsurface drip irrigation system described as under:

- ☆ Risk of clogging
- ☆ When saline water is used, salts accumulate at the wetting front
- ☆ Emitter can be damaged or blocked by root hairs
- ☆ Bacterial slimes and algae growing on the interior walls of the laterals and emitters combined with clay particles in the water can block the emitters

☆ Suspended organic matter and clay particles can damage the system

☆ A lot of repair work is caused by rodents chewing the tubes and Heavy machinery can damage the laterals

Microirrigation

☆ Microirrigation is the slow application of continuous drips, tiny streams or miniature sprays of water above or below the soil surface.

☆ Microirrigation system is effective in saving water and increasing water use efficiency as compared to the conventional surface irrigation method. Besides, it helps reduce water consumption, growth of unwanted plants (weeds), soil erosion and cost of cultivation.

Field capacity is the moisture or water content present in the soil after excess water has drained away and the rate of downward movement has decreased, which takes place within 2–3 days after a spell of rain or irrigation. It means that after drainage stops, the large soil pores are filled with both air and water, while the smaller ones are still filled with water. At this stage, the soil is said to be at field capacity and is considered to be ideal for crop growth.

☆ Microirrigation can be adopted in all kinds of land, especially where it is not possible to effectively use flooding method for irrigation. In flooding method of irrigation, a field is flooded with water. This results in significant run-off, anaerobic conditions in the soil and around the root zone, and deep irrigation below the root zone, which does not supply sufficient water to the plants. It is, therefore, one of the most inefficient surface irrigation methods.

☆ Microirrigation can be useful in undulating terrain, rolling topography, hilly areas, barren land and areas having shallow soils. According to depth, soil types can be classified as shallow (depth less than 22.5 cm), medium deep (22.5–45 cm) and deep soil (more than 45 cm).

Characteristics of Microirrigation System

☆ Water is applied via pressurised piping system. Microirrigation requires pumps for developing the required pressure for delivering water through pipelines, regardless of whether the source of water is surface or underground.

☆ Water is applied drop-by-drop for a long period in case of drip irrigation system.

☆ Water is applied at a low rate to maintain the optimum air–water balance within the root zone.

☆ Water is applied at frequent intervals as per the requirement of plants.

☆ Water is supplied directly to the plants and not to the other areas of the field, thus, reducing wastage.

☆ Soil moisture content is always maintained at 'field capacity' of the soil. Hence, crops grow at a faster rate, consistently and uniformly.

Drainage

☆ Agricultural drainage is the removal of excess water, known as free water or gravitational water, from the surface or below the farm so as to create favourable soil condition for plant growth. The lowering down of the ground water table (G.W.T.) below the root zone of the crops to improve the plant growth or to reduce the accumulation of salt is included in drainage.

☆ Irrigation and drainage are equally important for successful crop production. Irrigation and drainage cannot be separated from one another and they should 80 hand to hand. Excess water in the soil even for a short period may cause severe damage to crop and soil. Irrigation provides sufficient moisture in the soil for satisfactory growth of plant and drainage is important for avoiding excess moisture in the root zone.

☆ The excess moisture may originated from excess rainfall, over irrigation, Seepage from canal or reservoirs or ditches. A land having high water table or water stands on the *land* surface for a long period, excessive soil moisture content, humid or superhumid regions with continuous or intermittent rainfall or flat land with fine textured soil may need drainage for high agricultural productivity.

Types of Drainage

Following are the two types of drainage:

1. **Surface Drainage :** The process of removing the excess water from the land surface is known as surface drainage. Adequate arrangement for surface drainage in heavy rainfall areas is essential for speedy disposal of water. Rainfall, snowmelt, waste run off, seepage from adjoining higher land, overflow from stream channel *etc.* are the sources of surface water. Surface drainage system may be required in either humid or in irrigated areas which is usually an integral part of irrigation system on slowly permeable soils or in areas of high precipitation rates. It is therefore told that irrigation and drainage are like the two faces of a coin.

2. **Sub-Surface or Internal Drainages:** The process of removing the water from sub-surface or soil profile is known as sub-surface or internal drainage. There should be sub-surface drainage in high water table areas.

In either case, the system is conveniently divided into the following three functional parts.

☆ **Collection System:** Budding, surface field ditches, row ditches or diversion ditches are the integral part of the system that first picks up water from the land.

☆ **Disposal System:** This is a part of the system that receives water from the collection system and conveys it, usually in an open channel to the outlet.

☆ **Outlet:** This is the end point of the drainage system. The first step in the design of any drainage system is to locate the outlet, because most of the drainage systems fail due to failure of outlet. The most economic outlet is a natural water way, such as rivers, streams, ditches, municipal drains *etc.*

Importance of Drainage in Crop Production

Drainage and irrigation are equally important for successful crop production. Drainage, if not properly administered, causes various troubles in crop production, as follows:

☆ Soil aeration is hampered which ultimately effect the respiration process of plant root. As a result, the plant may die due to the damage of plant root.

☆ Soil becomes deficient in oxygen. The soil organism fails to survive in such soil. As a result, some poisonous substances are developed due to improper decomposition of organic matter which hamper the growth of plants.

☆ Loss of nutrient may occur due to leaching and run–off of excess water.

☆ Except some crops, most of the crops are very sensitive to water logging condition.

☆ The accumulated drainage water may be used for irrigation purposes by lifting them with suitable water lifting device.

☆ Soil erosion may occur. Surface salt accumulation may be the result of improper drainage.

☆ Sometimes drainage becomes a great problem than irrigation, especially in low lying areas.

☆ Surface salt accumulation may be the result of improper drainage.

☆ Drainage water of one field may be used as irrigation water with proper planning and designing.

Objectives of Drainage

There are some objects of drainage:

☆ Removal of excess water from the soil.

☆ Arrangement for good aeration in the soil.

☆ To control the loss of plant nutrients.

☆ To enhance the activity of soil organism.

☆ To improve the soil structure.

☆ To control soil erosion.

☆ To prevent the salt accumulation of surface soil.

☆ To develop favourable condition for the growth of plant root.

☆ To lower down the water level below the root zone.

Methods of Drainage

There are generally two methods that are employed for drainage of crop fields:

I.   Open Drainage

The open drains are easy to construct and maintain. The open drains are generally employed for drainage of surface water in our country. Open drains are being increasingly uses. Open drains are very suitable for an area where rainfall is heavy and drainage of more water is needed within a short period or time. Drainage channels are laid on the lowest contours of the land and they are utilised for draining of excess water from the land. The soils, having impervious lower layer, are best suited for open drainage.

Disadvantages

This system have the following disadvantages:

☆ Wastage of land is the main disadvantage of this method as the drains occupy considerable land.

☆ Drains within the land hamper the cultivation practices.

☆ Soil erosion is more.

☆ The drains need repairing every year. As a result, cost of cultivation is increased.

☆ Weeds grow on the side of drainage channels and their seeds are disposed of every where.

☆ Drainage of subsurface water is not possible.

II. Closed Drainage

The closed drains are laid underground. The closed drains are desirable for a place where the land is costly. The closed drains save the land, but it is very costly. Generally tile drains, mole drains, bush stone drains *etc.* are used as sub-surface drains.

(1) Tile Drains

The tile made of clay or concrete pipe or perforated steel pipes are mostly used for tile drains. The tile is about 30 to 50 cm. in length and 7 to 12cm. in diameter. The tiles are installed end to end in the field after digging trenches with 2–3 millimetre spacing in the joint. The drains are usually spaced 50–150 feet apart according to the type of soil. Excess water enters the system through the space in between the two tiles and conveyed along the gradient (Dastane, 1980).

The tiles are placed in underground soil in the following two methods:

Herring Bone or Fishbone Method

Herring bone pattern is used if the above submain is laid in a depression. The laterals join from each side alternatively and collects the water from eitherside.

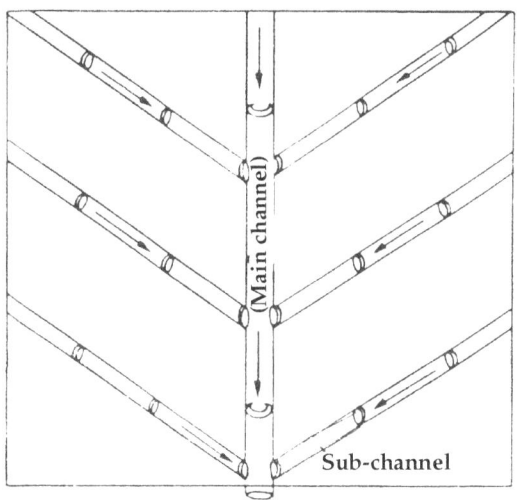

Herring Bone (Fishbone) Method.

Grid Iron Method

The grid iron method must be used where the land is practically level. It is a good method when the entire area is to be drained. Laterals enter the submain one side only to minimize double drainage.

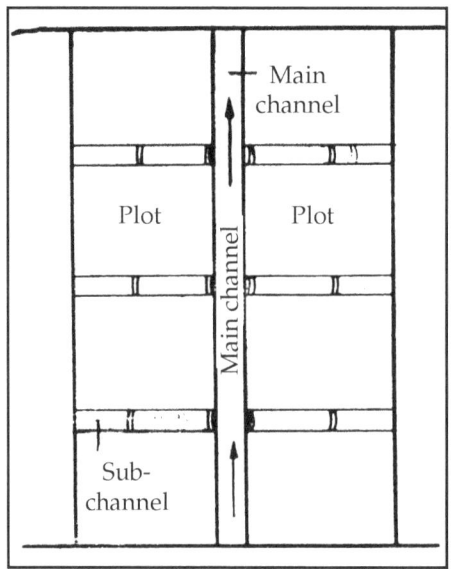

Grid Iron Method.

Distance and Depth of Drains

The distance and depth of drains vary with the types of soil as follows:

Soil Type	Distance of the Drains	Depth of the Drains
Sandy-loam soil	9–12 metre	90–120 cm
Loamy soil	6–9 metre	75–90 cm
Clay soil	3.5–6 metre	60–75 cm
Peat soil	5.5–6.3 metre	100–120 cm

(2) Mole Drains

- ☆ Mole drains are unlined, circular earthen channels in the subsoil made with a tractor drawn mole plough. The channels are made 3.0–3.5 metres apart.
- ☆ The channels receive and discharge the water to drainage ditches at the edge of the field. Mole drains are suitable only on heavy clay soils.

(3) Bush and Stone Drains

- ☆ Trenches are made in the field and they are filled up with bush and stone. Water may pass through the open space of bush and stone.
- ☆ Afterwards, the bushes are dried and decomposed and made into a fine channel. Drainage is continued until the drains are closed with soil.
- ☆ There is every possibility of closing the drains with soil.

Drainage system depends on the following factors:

☆ Topography and gradient of the land.

☆ Types of the soil.

☆ Permeability characteristics of the soil.

☆ Crop and its water tolerance capacity.

☆ The quantity of water to be disposed and its speed.

Chapter 8

Crop Rotation

Introduction

☆ Crop rotation is the practice of planting different crops sequentially on the same plot of land to improve soil health, optimize nutrients in the soil, and combat pest and weed pressure.

Crop rotations are planned sequences of different crops on the same field over time. Rotating crops provides productivity benefits by improving soil nutrient levels and breaking crop pest cycles. Farmers may also choose to rotate crops to reduce their production risk through diversification or to manage scarce resources such as labor during planting and harvesting time.

☆ For example, say a farmer has planted a field of corn. When the corn harvest is finished, he might plant beans, since corn consumes a lot of nitrogen and beans return nitrogen to the soil.

☆ A basic principle of crop rotation is not to grow the same thing in the same place two years running. In fact, the larger the gap between a crop occupying the same piece of ground the better. Some pests may be present at high levels initially but a gap of three or four years will see their numbers fall to acceptable levels without a host to sustain them.

☆ There are many different systems for rotating crops, some fairly crude and some quite complex, designed to ensure that following crops utilize nutrients left by previous crops. The simplest is a three year crop rotation but a four year crop rotation plan or even a five year crop rotation may suit you better.

Why Follow Crop Rotation?

☆ If you grow the same crop in the same place year after year you will get a buildup of pests and diseases specific to that crop. Different crops take different levels of nutrients from the soil and inevitably these become unbalanced, exhausting one nutrient but leaving a lot of another. This is often referred to as a 'sick soil'. Even the addition of fertilizers is unlikely to help since it is likely the trace elements and micro-nutrients are depleted in the same way.

☆ Some gardeners persist in growing their runner beans or onions in the same place each year but it has been proven this is not a good idea – not every old fashioned method is good! Rotating crops will reduce losses to pest and disease. Combine this with better use of nutrients and you will find increased yields from the same area of land.

Advantages of Crop Rotation

☆ Crop rotation helps return nutrients to the soil without synthetic inputs. The practice also works to interrupt pest and disease cycles, improve soil health by increasing biomass from different crops' root structures, and increase biodiversity on the farm. Life in the soil thrives on variety, and beneficial insects and pollinators are attracted to the variety above ground, too.

☆ Crop rotation is the practice of growing different crops in the same field over several seasons. To be precise, this means rotating between crops and fallow, a plowed, harrowed field left for a season without being sown. Farmers consider different factors when planning their crop rotation: the soil type, climate, amount of precipitation, herbicide residues, and market demand for agriculture products.

☆ Pest and disease protection: By rotating crops in a field, you deprive parasites of their habitual environment, thereby preventing crop diseases. The reverse is also possible. If crop rotation isn't done correctly, you can infect your next crop with diseases from the previous one. For example, the beet cyst eelworm (*Heterodera schachtii*) that infects sugar beets also affects rapeseed. That means that farmers should avoid planting these crops in the same field one after the other.

☆ When planning crop rotation, keep in mind the residual effects of the herbicides you used to treat previous crops. For example, the residual effect of sulfonylurea herbicides can adversely affect sugar beets, while triazine herbicides may negatively impact alfalfa seedlings.

☆ Maintaining soil fertility: If you don't rotate crops, the soil in that field will inevitably begin to lose the nutrients plants need to grow. You can avoid this by sowing crops that increase organic matter and nitrogen in the soil.

☆ Some farmers, for instance, sow legumes to take advantage of their symbiotic relationship with Rhizobia bacteria found in the soil. Why? When coming into contact with legume roots, Rhizobia bacteria form nodules where they convert nitrogen into ammonia, which the plant uses to grow. Likewise, crucifer green manures, such as white mustard, heal the soil by releasing substances that suppress the *Rhizoctonia solani* fungus.

☆ Alternating between crops with different root systems also offers benefits. Plants with longer roots can get nutrients from deeper layers of the soil than those with a shorter root system. When you alternate between crops like this, you keep the soil healthy.

☆ And yet, even farmers that know the benefits of crop rotation still sometimes don't do it. Say that the price for corn is high for several years in a row. The temptation arises to sow the fields with it nonstop to cash in while prices are high. In the long run, however, this practice leads to problems. Soil fertility decreases and it becomes extremely difficult to grow any crop with high yields.

Principles of Crop Rotation

☆ The crops with tap roots should be followed by those with fibrous root system. This helps in proper and uniform use of nutrients from the soil.

☆ Leguminous plants have the ability to fix atmospheric nitrogen to form nitrogen compounds through the help of certain bacteria present in their root nodules. These nitrogen compounds go into the soil and make it more fertile.

☆ When the cereal crops like rice, wheat, maize are grown in the soil, it uses up a lot of nitrogenous salts from the soil. If another crop of cereal is grown in the same soil, the soil becomes nitrogen deficient.

☆ The leguminous crops should be grown after non-leguminous crops. Legumes fix atmospheric nitrogen in the soil and add more organic matter to the soil.

☆ In the rotation of crops, leguminous crops like pulses, beans, peas, groundnut and Bengal gram are sown after cereal crops like wheat, maize and pearl millet.

☆ The leguminous plants are grown alternately with non-leguminous plants to restore the fertility of the soil.

☆ The crops which uptake lesser amount of nutrients from the soil are called less exhaustive crops like oilseeds. The crops which uptake higher amount of nutrients from the soil are called exhaustive crops like potato, sugarcane.

☆ The exhaustive crops are grown alternately with less exhaustive crops to restore the fertility of the soil and Therefore, The exhaustive crops are grown alternately with less exhaustive crops to restore the fertility of the soil.

☆ Continuous cultivation of exhaustive crops in the same land leads to gradual reduction of soil fertility or nutrient status of the soil. More exhaustive crops should be followed by less exhaustive crops.

☆ The crop of the same family should not be grown in succession because they act like alternate hosts for pests and diseases.

☆ An ideal crop rotation is one which provides maximum employment to the family and farm labour, farm machineries and equipments are efficiently used.

☆ Selection of the crop should be demand based.

☆ The selection of crops should be problem based.

☆ The selection of crops should suit to the farmer's financial conditions.

☆ The crops selected should also suit to the soil and climate conditions.

☆ The rotation should provide roughage and pasturage for the livestock kept on farm. Therefore, fodder crop must be included in the rotation to feed the animals. Examples: rice- black gram-fodder maize/berseem

☆ On sloppy lands alternate cropping of crosion promoting and erosion resisting crops should be adopted. Examples: Maize (erosion Promoting) - cow pea/green gram (erosion resisting).

☆ Under Dryland or limited irrigation condition, drought tolerant crops (Jowar, Bajra), in low lying and flood prone areas, water stagnation tolerant crops (Paddy, Jute) should be adopted.

☆ The crops of the same family should not be grown in succession because they act like alternate hosts for insect pests and disease pathogens and weeds associated with crops.

☆ The selection of crops should be according to the need/demand base. That is according to need of the people of the area and family. Example: Vegetables and flowers are grown in areas closer to the cities for higher income.

☆ The sequence of cropping adopted for any specific area should be based on proper land utilization.

☆ Rotation of crops helps in weed control and pest control. This is because weeds and pests are very choosy about the host crop plant, which they attack. When the crop is changed the cycle is broken. Hence, pesticide cost is reduced.

☆ Rotation of crops helps in saving on nitrogenous fertilizers, because leguminous plants grown during the rotation of crops can fix atmospheric nitrogen in the soil with the help of nitrogen fixing bacteria.

☆ There is an overall increase in the yield of crops due to maintenance of proper physical condition of the soil and its OM content.

4.4 FACTORS AFFECTING CROP ROTATION

☆ Climate: Climate is the one of most important factor which is effect the crop rotation either by wind, rain or other factors.

☆ Type and nature of soil: Type and nature of soil is also important factor which effects the crop rotation some soil are fertile and some are low in fertility.

☆ Availability of inputs: Availability of inputs at the place is also effects the crop rotation like fertilizer, pesticide *etc.*

☆ Availability of labor: Availability of labor is effect the crop rotation. The labor is required at the critical stages of crop if the labor is not available at that time the crop may cause loss.

☆ Situation of farm: The farm location is also very important factor which is effect the crop rotation.

☆ Size of Farm: The size of farm is effects the crop rotation. Small land holding is major problem in Pakistan that's why crop rotation is effect by the farm size.

Benefits of Crop Rotation

☆ Agronomists describe the benefits to yield in rotated crops as "The Rotation Effect". There are many benefits of rotation systems. The factors related to the increase are broadly due to alleviation of the negative factors of monoculture cropping systems. Specifically, improved nutrition; pest, pathogen, and weed stress reduction; and improved soil structure have been found in some cases to be correlated to beneficial rotation effects.

☆ Other benefits of rotation cropping systems include production cost advantages. Overall financial risks are more widely distributed over more diverse production of crops and/or livestock. Less reliance is placed on purchased inputs and over time crops can maintain production goals with fewer inputs. This in tandem with greater short and long term yields makes rotation a powerful tool for improving agricultural systems.

Soil Organic Matter

☆ The use of different species in rotation allows for increased soil organic matter, greater soil structure, and improvement of the chemical and biological soil environment for crops. With more SOM, water infiltration and retention improves, providing increased drought tolerance and decreased erosion.

☆ Soil organic matter is a mix of decaying material from biomass with active microorganisms. Crop rotation, by nature, increases exposure to biomass from sod, green manure, and a various other plant debris. The reduced need for intensive tillage under crop rotation allows biomass aggregation to lead to greater nutrient retention and utilization, decreasing the need for added nutrients. With tillage, disruption and oxidation of soil creates a less conducive environment for diversity and proliferation of microorganisms in the soil. These microorganisms are what make nutrients available to plants. So, where active soil organic matter is a key to productive soil, soil with low microbial activity provides significantly fewer nutrients to plants; this is true even though the quantity of biomass left in the soil may be the same.

☆ Soil microorganisms also decrease pathogen and pest activity through competition. In addition, plants produce root exudates and other chemicals which manipulate their soil environment as well as their weed environment. Thus rotation allows increased yields from nutrient availability but also alleviation of allelopathy and competitive weed environments.

Carbon Sequestration

☆ It is reported that crop rotations greatly increase soil organic carbon (SOC) content, the main constituent of soil organic matter. Carbon, along with hydrogen and oxygen, is a macronutrient for plants. Highly diverse rotations spanning long periods of time have shown to be even more effective in increasing SOC, while soil disturbances (*e.g.* from tillage) are responsible for exponential decline in SOC levels.

☆ In Brazil, conversion to no-till methods combined with intensive crop rotations has been shown an SOC sequestration rate of 0.41 tonnes per hectare per year. In addition to enhancing crop productivity, sequestration of atmospheric carbon has great implications in reducing rates of climate change by removing carbon dioxide from the air.

Nitrogen Fixation

☆ Rotating crops adds nutrients to the soil. Legumes, plants of the family Fabaceae, for instance, have nodules on their roots which contain nitrogen-fixing bacteria called rhizobia. During a process called nodulation, the rhizobia bacteria use nutrients and water provided by the plant to convert atmospheric nitrogen into ammonia, which is then converted into an organic compound that the plant can use as its nitrogen source.

☆ It therefore makes good sense agriculturally to alternate them with cereals (family Poaceae) and other plants that require nitrates. How much nitrogen made available to the plants depends on factors such as the kind of legume, the effectiveness of rhizobia bacteria, soil conditions, and the availability of elements necessary for plant food.

Control of Pest and Diseases

☆ Crop rotation is also used to control pests and diseases that can become established in the soil over time. The changing of crops in a sequence decreases the population level of pests by interrupting pest life cycles and interrupting pest habitat. Plants within the same taxonomic family tend to have similar pests and pathogens. By regularly changing crops and keeping the soil occupied by cover crops instead of lying fallow, pest cycles can be broken or limited, especially cycles that benefit from overwintering in residue.

☆ For example, root-knot nematode is a serious problem for some plants in warm climates and sandy soils, where it slowly builds up to high levels in the soil, and can severely damage plant productivity by cutting off circulation from the plant roots. Growing a crop that is not a host for root-knot nematode for one season greatly reduces the level of the nematode in the soil, thus making it possible to grow a susceptible crop the following season without needing soil fumigation. This principle is of particular use in organic farming, where pest control must be achieved without synthetic pesticides.

Weed Management

☆ Integrating certain crops, especially cover crops, into crop rotations is of particular value to weed management. These crops crowd out weed through competition. In addition, the sod and compost from cover crops and green manure slows the growth of what weeds are still able to make it through the soil, giving the crops further competitive advantage. By slowing the growth and proliferation of weeds while cover crops are cultivated, farmers greatly reduce the presence of weeds for future crops, including shallow rooted and row crops, which are less resistant to weeds. Cover crops are, therefore, considered conservation crops because they protect otherwise fallow land from becoming overrun with weeds.

☆ This system has advantages over other common practices for weeds management, such as tillage. Tillage is meant to inhibit growth of weeds by overturning the soil; however, this has a countering effect of exposing weed seeds that may have gotten buried and burying valuable crop seeds. Under crop rotation, the number of viable seeds in the soil is reduced through the reduction of the weed population.

☆ In addition to their negative impact on crop quality and yield, weeds can slow down the harvesting process. Weeds make farmers less efficient when harvesting, because weeds like bindweeds, and knotgrass, can become tangled in the equipment, resulting in a stop-and-go type of harvest.

Preventing Soil Lost from Erosion by Water

☆ Crop rotation can significantly reduce the amount of soil lost from erosion by water. In areas that are highly susceptible to erosion, farm management practices such as zero and reduced tillage can be supplemented with specific crop rotation methods to reduce raindrop impact, sediment detachment, sediment transport, surface runoff, and soil loss.

☆ Protection against soil loss is maximized with rotation methods that leave the greatest mass of crop stubble (plant residue left after harvest) on top of the soil. Stubble cover in contact with the soil minimizes erosion from water by reducing overland flow velocity, stream power, and thus the ability of the water to detach and transport sediment. Soil Erosion and Cill prevent the disruption and detachment of soil aggregates that cause macropores to block, infiltration to decline, and runoff to increase. This significantly improves the resilience of soils when subjected to periods of erosion and stress.

☆ When a forage crop breaks down, binding products are formed that act like an adhesive on the soil, which makes particles stick together, and form aggregates. The formation of soil aggregates is important for erosion control, as they are better able to resist raindrop impact, and water erosion. Soil aggregates also reduce wind erosion, because they are larger particles, and are more resistant to abrasion through tillage practices.

☆ The effect of crop rotation on erosion control varies by climate. In regions under relatively consistent climate conditions, where annual rainfall and temperature levels are assumed, rigid crop rotations can produce sufficient plant growth and soil cover. In regions where climate conditions are less predictable, and unexpected periods of rain and drought may occur, a more flexible approach for soil cover by crop rotation is necessary. An opportunity cropping system promotes adequate soil cover under these erratic climate conditions. In an opportunity cropping system, crops are grown when soil water is adequate and there is a reliable sowing window. This form of cropping system is likely to produce better soil cover than a rigid crop rotation because crops are only sown under optimal conditions, whereas rigid systems are not necessarily sown in the best conditions available.

☆ Crop rotations also affect the timing and length of when a field is subject to fallow. This is very important because depending on a particular region's climate, a field could be the most vulnerable to erosion when it is under fallow. Efficient fallow management is an essential part of reducing erosion in a crop rotation system.

☆ Zero tillage is a fundamental management practice that promotes crop stubble retention under longer unplanned fallows when crops cannot be planted. Such management practices that succeed in retaining suitable soil cover in areas under fallow will ultimately reduce soil loss. In a recent study that lasted a decade, it was found that a common winter cover crop after potato harvest such as fall rye can reduce soil run-off by as much as 43 per cent, and this is typically the most nutritional soil.

Increasing the Biodiversity

☆ Increasing the biodiversity of crops has beneficial effects on the surrounding ecosystem and can host a greater diversity of fauna, insects, and beneficial microorganisms in the soil.

☆ Some studies point to increased nutrient availability from crop rotation under organic systems compared to conventional practices as organic practices are less likely to inhibit of beneficial microbes in soil organic matter, such as arbuscular mycorrhizae, which increase nutrient uptake in plants. Increasing biodiversity also increases the resilience of agro-ecological systems.

Farm Productivity and Risk Management

☆ Crop rotation contributes to increased yields through improved soil nutrition. By requiring planting and harvesting of different crops at different times, more land can be farmed with the same amount of machinery and labour.

☆ Different crops in the rotation can reduce the risks of adverse weather for the individual farmer.

Reduced Usage of Pesticides

☆ Certain species are attacked by certain pests, for example, potatoes are cut up by Colorado beetles. They are killed with target chemicals. When they are used for many years, excessive amounts pollute nature being harmful to all living beings.

☆ For example corn or wheat, they will leave the field, as they simply don't eat that. At the same time, these insects are quite comfortable with tomatoes or aubergines – so this change won't solve the problem.

Nature Protection

☆ The chemical form of nitrogen pollutes soils and waters. Besides, plants absorb but a small part of nitrogen from fertilizers, the rest harms our ecology.

Limitations of Crop Rotation

☆ Specialization in one crop is not possible.

☆ Requirement of equipments and machineries varies from crop to crop.

☆ Allelopathic effect of preceding crop.

☆ Serves as alternate hosts for pests and diseases.

☆ Many challenges exist within the practices associated with crop rotation. For example, green manure from legumes can lead to an invasion of snails or slugs and the decay from green manure can occasionally suppress the growth of other crops.

Objectives of Crop Rotation

☆ To prevent the built up of insect pest, weeds and soil born diseases

☆ To maintain soil fertility for the next crop

☆ To conserve soil erosion which may cause from wind or water

☆ To conserve soil moisture from one season for the next

☆ To ensure a balanced programme of work throughout the season.

Chapter 9

Allelopathy

Introduction

☆ The term allelopathy was coined by Prof. Hans Molisch in 1937 which indicates stimulatory/inhibitory biochemical interactions between the plants including microorganisms. Allelopathy is a biological phenomenon by which an organism produces one or more biochemicals that influence the growth, survival, and reproduction of other organisms. These biochemicals are known as allelochemicals and can have beneficial (positive allelopathy) or detrimental (negative allelopathy) effects on the target organisms. Allelochemicals are a subset of secondary metabolites, which are not required for metabolism (*i.e.* growth, development and reproduction) of the allelopathic organism. Parthenium daughter plants exhibiting teletoxy to its parent plants is known as autotoxy. The word allelopathy is derived from Greek – allelo, meaning each other and patho, an expression of sufferance of disease.

☆ Allelopathy is characteristic of certain plants, algae, bacteria, coral, and fungi. Allelopathic interactions are an important factor in determining species distribution and abundance within plant communities, and are also thought to be important in the success of many invasive plants. Allelochemicals are found to be released to environment in appreciable quantities via root exudates, leaf leachates, roots and other degrading plant residues, which include a wide range of phenolic acids such as benzoic (1) and cinnamic acids (2), alkaloids (3), terpenoids (4) and others. These compounds are known to modify growth, development of plants, including germination and early seedling growth.

History of Allelopathy

✰ For over 2000 years, allelopathy has been reported in the literature with respect to plant interference. Ancient literature has described the growth of crops which "rob the soil of nutrients", "sicken the soil", and even referred to roots producing toxins which suppress plant growth. The earliest recorded observations of weed and crop allelopathy were made by none other than Theophrastus (300 BC) and Pliny II (1 AD).

✰ Pliny reported that chickpea (*Cicer arietinum*), barley (*Hordeum vulgare*), fenugreek (*Trigonella foenum-graecum*), and bitter vetch (*Vicia ervilia*) destroy or burn up farmland (Rice, 1984). He also described the shade of black walnut (*Juglans* spp.) as heavy and believed that walnut and its residues could cause potential injury to man and anything planted in the vicinity. Pliny was apparently aware that release of chemicals by plants contributed to this soil sickness; as evidenced by his statement *"The nature of some plants though not actually deadly is injurious owing to its blend of scents or of juice."*

✰ In the 1600's several naturalists noted in the English literature that certain plants do not grow well in the presence of each other. The Japanese literature also shows examples of plants causing injury to others due to the production of extracts of toxic compounds with rainfall, specifically Japanese red pine (*Pinus densiflora*) (Rice 1984).

✰ In the 1800's, agronomists started to note problems with repeated cropping of certain perennials. For example, Young (1804) discovered that clover was apt to fail in some regions of England where it is cultivated constantly due to soil sickness which accrues over time.

✰ De Candolle was one of the first to actually perform experiments examining the toxicity associated with root exudates (de Candolle 1832; Singh *et al.,* 2001).

✰ In 1881, Stickney and Hoy observed that vegetation under black walnut was very sparse in pasture settings, and pointed out that this might be due to high mineral requirement of the tree, or the poisonous character of the moisture dripping off the tree itself.

✰ As Rice indicates, if one peruses older literature, there are many examples described by botanists, farmers and gardeners that strongly suggest allelopathic interactions among plants. It is interesting to note that many of the species demonstrated to have powerful medicinal effects on humans also have been subsequently demonstrated to have powerful allelopathic effects as well (Chevallier 1996; Rice 1984; Wink 1999).

✰ Interest in the field of allelopathy revived again in the 20th century, with the development of suitable techniques for extraction, bioassay, and chemical isolation and identification (Willis 1997).

☆ Mc Calla *et al.,* published a series of papers from 1948 until 1965 that described allelochemicals produced from plant residues and the importance of the interaction of microbes upon the decomposition of these residues (Putnam and Weston 1986).

☆ Two outstanding contributors to the field of allelopathy during this period also included C.H. Muller and his associates at Santa Barbara, California who published many articles on volatile inhibitors produced by plants growing in the chaparral and desert. E.L. Rice at the University of Oklahoma contributed many papers to the field and described impacts of allelochemicals on nitrifying and nitrogen fixing bacteria in the soil rhizosphere, as well as classical works documenting allelopathy in prairie type ecosystems of the central U. S. (Bell and Muller 1973; Muller 1969; Ric, 1984; Putnam and Weston 1986).

☆ Rick Willis is now finishing an exhaustive review that thoroughly documents the history and the science of allelopathy as a field of research. Some of his work has recently been published in the Allelopathy Journal (Willis 1997; Willis 2000). The number of publications in the field of allelopathy has increased exponentially as physiologists, soil scientists, weed scientists and natural products chemists continue to study this challenging area (Macias 2002).

Allelochemicals are released in the form of:

☆ Vapour (released from plants as vapour): Some weeds release volatile compounds from their leaves. Plants belonging to labiateae, compositeae yield volatile substances.

☆ Leachates (from the foliage): From Eucalyptus allelo chemicals are leached out as water toxins from the above ground parts by the action of rain, dew or fog.

☆ Exudates from roots: Metabolites are released from Cirsium arvense roots in surrounding rhizosphere.

☆ Decomposition products of dead plant tissues and warn out tissues.

Types of Allelopathy

Weed on Crop

☆ *Agropyron repens* (Quack grass) is an important weed of field crops, which causes serious decreases in yield of maize and potato. It interferes with uptake of manures, particularly nitrogen and potassium by maize. Ethylene is generated in quack grass rhizomes due to microbial activity in soil, which is responsible for allelopathic effects of the weed resulting in decrease uptake of mineral by associated crops.

☆ *Avena fatua* (Wild oat) is a serious weed of winter annuals like wheat, barley and oats. Wild oat residues inhibit germination of certain herbaceous species in shrubs stand due to an allelopathic mechanism. Growth of leaves and roots of wheat is significantly reduces by root exudate of wild oat.

☆ *Cynodon dactylon* (Bermuda grass) found on cultivated lands. Decayed Bermuda grass residues remain in the field inhibits seed germination, root and top growth of barley due to allelopathic effect.

☆ *Cyperus esculentus* (Yellow Nut sedge) is a perennial nut sedge infesting grain crops, soybean, orchards *etc.* It inhibits root and shoot growth of maize and soybean. The effect of soybean is due to the allelopathic compounds- vanillic acid, hydroxybenzoic acid in the yellow nut sedge extract.

☆ *Sorghum halepense* (Johnson grass) is a persistent perennial weed in sugarcane, maize, soybean *etc.*? Root exudates and decaying residues of Johnson grass can inhibit both root and shoot growth.

☆ *Setaria viridis* (Giant foxtail) – Yield reduction in corn is due to the inhibitory effect of exudates of mature giant foxtail roots and leachates of dead roots.

☆ *Impereta cylindrica* (Cogon grass) inhibits the growth tomato and cucumber.

☆ Field bindweed, Canada Thistle- release root exudates that affect seedling growth of many crops *e.g.* cabbage, carrot, tomato, radish *etc.*

Weed on Weed

☆ *Impereta cylindrica* (Cogon grass)– inhibits the emergence and growth of an annual broadleaf weed *i.e. Borreria hispada* (Button weed) by exudating inhibitory substances through rhizomes.

☆ *Sorghum halepense* (Johnson grass) – living and decaying rhizomes and leaves inhibit the growth of *Setaria viridis* (Giant foxtail), *Digitaria sanguinalis* (Large crabgrass), and *Amaranthus spinosus* (Spiny amaranth).

Crop on Weed

☆ Coffea arabica (Coffee) release 1,3,7-trimethylxanthin which inhibts germination of *Amaranthus spinosus* (Spiny amaranth).

☆ *Zea mays* (Maize) root extracts increase catalase and peroxidase activity of the weeds which inhibit their growth.

☆ Oat, pea, wheat suppress the growth of Chenopodium album (Lamsquarter).

☆ Recently some rice genotypes have already been identified which have allelopathic effects on weeds.

☆ Allelopathic effect of crops and weeds on other weeds may be applied to develop natural herbicides.

Application of Allelopathy in Weed Management

Allelochemicals as Herbicides

☆ Allelochemicals offer excellent potential as herbicides. Firstly, they could be used directly as herbicides because these are free from all the problems associated with present herbicides. Secondly, their chemistry may be used to develop new herbicides.

☆ As traditional methods of discovering and developing new herbicides have become more difficult and expensive, the interest in natural products as sources of herbicide chemistry has increased. Besides, public awareness and demand for environmentally safer herbicides with less persistence, more specific targets and less potential for contaminating groundwater makes searches for new weed control strategies, using natural products more attractive.

☆ Plants and microorganisms produce hundreds of secondary compounds. Many of these compounds are phytotoxic and have potential as herbicides.

Allelopathic Chemicals

☆ Phenolic acid

☆ Coumarins – block mitosis in onion by forming multinucleate cells

☆ Terpienoids

☆ Flavinoids Scopulatens – inhibits photosynthesis without significant effect on respiration.

Allelopathic Control of Certain Weeds Using Botanicals

For instance Dry dodder powder has been found to inhibit the growth of water hyacinth and eventually kill the weed. Likewise carrot gross powder found to detrimental to other aquatic weeds. The presence of marigold (*Tagetes erecta*) plants exerted adverse allelopathic effect on *Parthenium* spp. growth. The weed coffeesena (*Cassia* sp.) show suppressive effect on Parthenium. The eucalyptus tree leaf leachates have been shown to suppress the growth of nut sedge and bermuda grass. Allelo chemicals are produced by plants as end products, by-products and metabolites liberalized from the plants.

I.   Allelopathic Effects of Weeds on Crop Plants

☆ Root exudates of Canada thistle (*Cirsium* sp.) injured oat plants in the field

☆ Root exudates of Euphorbia injured flax. But these compounds are identified as parahydroxy benzoic acid.

Maize

☆ Leaves and inflorescence of *Parthenium* sp. affect the germination and seedling growth

☆ Tubers of *Cyperus esculentus* affect the dry matter production

☆ Quack grass produced toxins through root, leaves and seeds interfered with uptake of nutrients by corn.

Sorghum

☆ Stem of Solanum affects germination and seedling growth

☆ Leaves and inflorescence of Parthenium affect germination and seedling growth.

Wheat

☆ Seeds of wild oat affect germination and early seedling growth

☆ Leaves of Parthenium affects general growth

☆ Tubers of *C. rotundus* affect dry matter production

☆ Green and dried leaves of *Argemone mexicana* affect germination and seedling growth

Sunflower

☆ Seeds of Datura affect germination and growth.

II. Effect of Weed on Another Weed

☆ Thatch grass (*Imperata cylindrica*) inhibited the emergence and growth of an annual broad leaf weed (*Borreria hispida*).

☆ Extract of leaf leachate of decaying leaves of Polygonum contains flavonoides which are toxic to germination, root and hypocotyls growth of weeds like *Amaranthus spinosus*

☆ Inhibitor secreted by decaying rhizomes of Sorghum halepense affect the growth of *Digitaria sanguinalis* and *Amaranthus* sp.

☆ In case of Parthenium, daughter plants have allelopathic effect on parent plant. This is called AUTOTOXY.

III. Effect of Crop on Weed

☆ Root exudates of wheat, oats and peas suppressed Chenopodium album. It increased catalase and peroxidase activity of weeds and inhibited their growth.

☆ Cold water extract of wheat straw reduces growth of Ipomea and Abutilon.

IV. Stimulatory Effect

☆ Root exudates of corn promoted the germination of Orbanchae minor; and *Striga hermonthica*. Kinetin exuded by roots sorghum stimulated the germination of seeds of *Stirga asisatica*.

☆ Strigol – stimulant for witch weed was identified in root exudates from cotton.

Allelopathic Effect at a Glance

Weeds	Crops	Cause/Source	Effect
Agropyron repens (Quack grass)	Maize, potato	Ethylene produced by the activity of microorganism on rhizomes	Decrease uptake of manures (N,K) followed by yield reduction
Avena fatua (Wilt oat)	Wheat, barley, oat	Root exudates	Growth of leaves and roots of wheat
Cynodon dactylon (Bermuda grass)	Barley	Decayed grass residues	Seed germination, root and top growth
Cyperus escdentus (Yellow Nut sedge)	Grain crops, soybean, orchard	Vanillic acid, Hydrobenzoic add in sedge extract	Root and shoot growth of maize and soybean
Sorghum halepense (Johnson grass)	Sugarcane, maize, soybeal	Root exudates and decaying residues	Root and shoot growth
Setalia viridis (Giant oxtail)	Maize	Roots and leachates of dead roots	Yield reduction
Impereta cylininca (Cogon grass)	Tomato and cucumber	Root extracts	Inhibit growth
Field bindweed. Canada Thistle	Cabbage. carrot, tomato etc.	Root exudates	Seedling growth
Impereta cylinitica (Cogon grass)	*Borreria hispada* (Button weed)	Exudates of inhibitory substances through rhizomes	Inhibits the emergence and growth
Sorghum halepense (Johnson grass)	*Setaria viridis* (Giant foxtail), *D. sanguinalis* (Large crabgrass)	Living and decaying rhizomes and leaves	Inhibit growth
Coffea arabica (Coffee)	*Amaranthus spinosus* (Spiny amaranth)	1,3,7-trimethylxanthin	Inhibit germination
Zea mays (Maize)	Associated weeds	Increased Catalase and Peroxidase activity by root extract	Inhibit growth
Oat, pea, wheat	*Chenopodium album* (Lamsquarter)	Root exudates	Suppress growth

Factors Affecting Allelopathic Effect

Allelopathic effects might also depend on a number of other factors that might be important in any given situation:

☆ *Varieties:* There can be a great deal of difference in the strength of allelopathic effects between different crop varieties.

☆ *Specificity:* There is a significant degree of specificity in allelopathic effects. Thus, a crop which is strongly allelopathic against one weed may show little or no effect against another.

☆ *Autotoxicity:* Allelopathic chemicals may not only suppress the growth of other plant species, they can also suppress the germination or growth of seeds and plants of the same species. Lucerne is particularly well known for this and has been well researched. The toxic effect of wheat straw on following wheat crops is also well known.

☆ *Crop on crop effects:* Residues from allelopathic crops can hinder germination and growth of following crops as well as weeds. A sufficient gap must be left before the following crop is sown. Larger seeded crops are affected less and transplants are not affected.

☆ *Environmental factors:* Several factors impact on the strength of the allelopathic effect. These include pests and disease and especially soil fertility. Low fertility increases the production of allelochemicals. After incorporation the aleopathic effect declines fastest in warm wet conditions and slowest in cold wet conditions.

Chapter 10

Weed Management

Weed: An Introduction

☆ The weed is a plant which grow at unwanted place or an unwanted plant, or plant which may be extremely useless or unwanted or poisonous/ noxious, excluding fungi, which may interfere with the needs of people. Thus, weed plants are out of place, unwanted, non- useful, often prolific and persistent, competitive harmful, even poisonous that interfere with agricultural operations, increase labour, add to cost and reduce yields. More than 30,000 species of weeds have been identified world-wide. These weeds cause losses to crops. Therefore, it is essential that the weeds should be eradicated from crop fields.

☆ A weed is a plant grown at a place where it is not desired. The difference between then crop plant and weed plant was told by Jethro-tull as: "crop plants meet the needs of meanwhile the weeds compete with the crop plants".

☆ A comprehensive and widely accepted definition of weed is "a weed is a plant growing out of place and time". The definition of weed does not indicate a particular group of plant kingdom as weedy. It can be said that all weeds are plants but all plants may not be weeds. Thus, the weed is a relative term *e.g. Cynodon dactylon* and *Cenchurs ciliaris* are valuable plant in Pastures but in crop field these are well recognized as troublesome weeds.

☆ Though various weed management practices can be adopted to suppress the weed, for effective and efficient weed management, a sound knowledge about the weed flora of that agro climatic zone is of prime importance.

Characteristics of Weeds

☆ Weeds are prolific, with abundant seed production and *Amaranthus* spp.

☆ Weeds remain dormant and viable for 30 to 40 years *e.g. Chenopodium* spp.

☆ Some weeds are propagated by vegetative part *e g. Cynodon* spp. *Convobuslus* spp.

☆ Some seeds of weeds are very similar to crop seeds and their separation becomes a problem and mustard and *Argemone* spp.

☆ Weeds are persistent and resistant to their control and eradication.

Loss in Crop Yields Due to Seeds

Crop	Reduction in Yields Due to Weeds (per cent)	Crop	Reduction in Yields Due to Weeds (per cent)
Rice	41.6	Groundnut	33.8
Wheat	16.0	Sugarcane	34.2
Maize	39.8	Sugar beet	70.3
Millets	29.5	Carrot	47.5
Soybean	30.5	Cotton	72.5
Gram	11.6	Onion	68.0
Pea	32.9	Potato	20.1

Important Weeds of some Common Crops and Herbicides Used

Crop Name	Important Weeds	Name or Recommended Weedicide(s)
Wheat	Genhusa (canary grass), Jangali jai (wild oats), Bathua (Lambs quarter), Hirankhuri (Bind weed), Satgathia (corn spury). Senji (sweet clover), Akers (common vetch), Matari (wild pea), Krishnaneel (scarlet).	Grassy weeds: Isoproturon 50 per cent @ 1.5 Kg/ha as post-emergence or Sulphosulphuron 75 per cent @ Broad leaved weeds: 2,4-D-(Na Salt) 80 per cent @ 625 gin as post-emergence or Met Sulfuron methyl 20 per cent @ 20 g: ha
Chickpea	Bathua (Lambs quarter), Pyazi (wild onion), Senji (sweet clover), Akers (vetch), Matari (wild pea), Khesari (*Lathyrus*), Titali (Titali), Krishnaneel (scarlet), Hirankhuri (Bind weed)	Pendimethalin 30 per cent ® 3.3 1/ha as pre-emergence or Fluchloralin 45 per cent @ 2.2 littha as pre-planing.
Field pea	Bathua (Lambs quarter), Pyazi (wild onion), Matari (wild pea), Krishnaneel (scarlet), Altars (vetch), Senji (sweet clover), Satgathia (corn spury), Khesari (*Lathyrus*)	Pendimethalin 30 per cent @ 3.3 1/ha as pre-emergence.

Crop Name	Important Weeds	Name or Recommended Weedicide(s)
Lentil	Bathua (Lambs quarter), Krishnaneel (scarlet), Pyazi (wild onion), Senji (sweet clover), Akera (vetch), Satgathia (corn spury)	Pendimethalin 30 per cent @ 3.3 1/ha as pre-emergence or Fluchloralin 45 per cent @ 2.21/ ha as pre-planting.
Sugarcane	Kans (Tiger grass), Mahakua (Agent runt), Motha (Nut grass), Johnson grass (Barru), Patherchata (carpet weed).	Atrazin @ 1.2 kg/ha or Ametrytte @ 1.5-2.0 kg/ha as pre-emergence.
Potato	Bathua (Lambs quarter), Kristmancel (scarlet), Senji (sweet clover), Satgathia (corn spury)	Atrazin 0.25 to 0.5 kg/ha. or 0.25 kg/ha Nietribuzin as pre-emergence.
Mustard	Bathua (Lambs quarter), Krishnancel (scarlet), Senji (sweet clover), Pyazi (wild onion), Satgathia (corn spury)	Fluchloralin 45 per cent @ 2,2 l/ha as pre-plant incorporation or Pendimethalin 30 per cent @ 3.3 l/ha as pre-emergence.
Safflower	Wild Safflower (*Carthamus oxycantha*)	
Sunflower	Wild sunflower (*Helianthus* spp.)	
Napier grass	Canada thistle (*Cirsiurn arvense* L. scop)	
	Dodder (*Cuscutta* spp. Baru)	
Johnson grass (*Sorghum halpense* L. pers)		
Quack grass (*Agropyron repens* L.P. Beauv)		
Hiran Khuri Wild morning glory (*Convulvulus arvensis* L.)		

Seed Characters of Some Common Weed Families

Family: Caryophyllaceae

☆ **Spergula arvensis:** The seed is 1-15mm diameter lens shaped, dull black, thin, flattish with wing. The embryo is linear, "U shaped without endosperm.

Family: Chenopodiaceae

☆ **Chenopodium album** (Bathuva): The seed is circular, flat and round; diameter 111/2mm, black, smooth, and shiny surface. Chenopodiu murale (Bathva): Similar to C. album but slightly bigger in size and dull in appearance.

Family: Convolvulaceae

☆ **Convolvulus arvensis** (Field weed): The seed colour is dull, grayish brown, length about 4 to 4 1/2mm; surface roughened with fine tubercles or short wavy lines. Back side convex and lateral plane, scar: inverted and #39;U and #39; shape and at right angles to the seed's long axis.

☆ **Ipomea hederacea:** The seed is diverse in shape (trigonous wedge, two inner faces are equal): size (lanceolate, ovoid to globose surface); smooth and colour: brown black. Scar: horse shoe shape and usually parallel to long axis.

Family: Poaceae

☆ *Phalaris minor* **Retz.** (Little seed canary grass, Canary grass, Gulli danda): Seeds are hard, with palet covering the grain, which is oval with an acute angle at one end and about 3- 5mm long, glossy and brownish grey in colour.

☆ *Avena fatua* (Wild oat): The seed consists of mature floret, narrowly cylindrical, tapering at apex, bears a twisted and bent dorsal awn, ventral side flat with fine grooves; colour: grey, brown or black, yellow to white.

☆ *Panicum* **spp:** The seed unit consists of one seeded spikelet. The grain surrounded by glumes (thin and papery). Lemma and Palea (hard, smooth and shiny, size: 1 1/2 to 2 ¾ mm usually lance shape.

☆ *Setaria etalica:* **Bandri Bandra.** The seed unit consists of one seeded spikelet. TheNgrain is surrounded by glumes (thin, papery and smooth). Lemma and Palea (hard, smooth and shiny)

Family: Liliaceae

☆ *Asphodelus tennuifolius* **(wild onion):** The seed is 1- 1/4 long, flattened eleptical three angled, (sharp) acute and black (crustraceous) testa.

Family: Papaeraceae

☆ *Argemone mexicana* **L (Maxican poppy, Satyanashi):** Seeds are nearly 2 mm long, ovoid, spherical surface with angular depression and a crest along one side, blackish brown in colour

☆ *Fumaria parviflora:* Fruit is very small, globose, one seeded, indehicent nutlet, rugose when dry and rounded at the top with two pits, color usually green.

Family: Papillionaceae

☆ *Medicago sativa* **(Lucerne):** The seed is roughly oval (scra lies in board indentation near one end or kindly shape twisted the congaxes (scar lies in middle of a distinct notch). Colour greenish yellow or light brown, length 1½ mm and width 2½ mm to 3mm.

☆ *Melilotus alba* **(white sweet clover):** The seed is identified by size (bigger length about 2 ½ mm and width 1½ mm), shape oblong to oval and translucent in appearance), and colour. (golden yellow to light brown). Scar lies in shallow indentation near top.

Family: Polygonaceae

☆ *Rumex* **sp (wild spinach):** Seed three sided acute as both ends, brown, spinning segments if present with long, fine teeth on the margins.

Family: Compositae

- ☆ *Helianthus* **spp (Wild Sunflower):** Seeds are, long in size, trigonous, very small hairs present on seed surface and dark brown to black in colour.

- ☆ *Cichorium intybus* **(Coffee chicory, Large rooted chicory, Chicory):** Seeds are up to 3 mm Long, trigonous wedge shaped, pale brown to gray and white in colour, pappus of scales present.

- ☆ *Carthamus oxycantha* **(Wild safflower):** Seeds are smaller than that of cultivated safflower, elongated in shape, grayish in colour with variegation/mottling on seed coat.

Classification of Weeds

There are many ways in which weeds can be classified. Two of the most common ways are by **gross morphological features,** and by their **lifecycle.** When classified by their gross morphological features, weeds are broken into three major categories: grasses, sedges, and broadleaf weeds. When classified by their life cycle, they are broken into annual, biennial, and perennial. The latter classification method can have a profound impact on the effectiveness of control measures.

Classifying Weeds by Gross Morphological Features

- ☆ Weeds considered **true grasses** are monocotyledons having only one seed-leaf when seedlings emerge from the soil. They have leaves that are long and narrow with a parallel venation pattern. Leaves arise in an alternate pattern on each side of a hollow stem or **culm**. Each leaf has two parts: the lower portion called the **sheath,** which is wrapped around the culm, and the upper portion called the **blade**. Most have fibrous root systems. Some examples of weeds that are grasses include crabgrass, sandbur, and dallisgrass.

- ☆ Weeds considered **sedges** are also monocotyledons, but not true grasses, though they exhibit a lot of the same characteristics as grasses. They are different from grasses in that their stems are solid, triangular, and have no nodes. Their leaves have a three-ranked arrangement (instead of an alternate arrangement) with each leaf one third the way around the stem from the one below it. The basal portion of each leaf forms a tube around the stem with no clear division between sheath and blade. Examples of weeds that are sedges are nutsedge and green kyllinga.

- ☆ **Broadleaf weeds** are different from grasses and sedges in that their leaf blades are expanded. Their leaves have netted venation as they are dicotyledons. Their stems branch as the plant grows, and they often have flowers. When broadleaf weed seedlings emerge from the soil, they have two seed-leaves. They often have tap roots or coarsely branched roots. Because of the diversity of dicots, it is important to first identify broadleaf weeds accurately before developing a management plan. Some common broadleaf weeds include dandelion, clover, burdock, and thistle.

Classifying Weeds by their Life Cycle

☆ **Annual weeds** complete their life cycle in one year or less. During that time, they germinate, complete their growth cycle, flower, produce seeds, and dia. Some develop prostrate stems or adventitious roots. If the stems are cut, they may develop into new plants. Annual weeds can be further divided based on the time of year their life cycle begins and ends.

❖ **Summer annual weeds** or **warm-season annual weeds** like crabgrass germinate in spring and die in fall when weather turns starts to turn cold. The seeds they produce during the summer growing season lay dormant in the soil during winter months and germinate the following spring.

❖ **Winter annual weeds** or **cool-season annual weeds** like henbit, annual bluegrass, and common chickweed germinate and develop in fall. They overwinter and mature the following spring. Then they flower, seed, and die during the summer.

☆ **Biennial weeds** have life cycles that span two years (*i.e.* two growing seasons). During the first year, biennials germinate, and the plant focuses on growth. The plant overwinters, and then during the second year or growing season, it flowers, produces seeds, and dies.

☆ **Perennial weeds** have life cycles that span more than two years. They reproduce from seeds or vegetative parts of the plant like rhizomes, bulbs, tubers, and stolons. Like annuals, perennial weeds can be classified into cool-season perennial weeds and warm-season perennials based on the time of year when they grow.

Types of Weeds

1. Annual, Biennial, and Perennial Weeds

☆ Annual weeds attain their full growth in one season, living for a few weeks, few months, or at the most for one year. Within this period they produce flowers and set seeds, and then die at the end of season. According to their major growth season these are called Kharif weeds and rabi weeds. *Chenopodium album, Phalaris minor, Ageratum conyzoides, Echinochloa* spp., and Digera arvensis are a few of the numerous annual weeds with us.

☆ While their regrowth is dependent each time upon seeds, a few annual weeds like Parthenium hysterophorus and Amaranthus spp. can give rise to new plants also from their crown buds. Also, P. hysterophorus and some other wasteland weeds may produce flowers and seeds round the year instead of at the end of a particular season.

☆ The main target in the control of annual weeds is to destroy these before their flowering period to prevent further seed production in them. Both, tillage and herbicides prove more effective against annual weeds when these are treated at their seedling stages than at their grown-up stages.

☆ Several soil active herbicides are now available to control weeds from seeds, before or just after their emergence from the soil.

☆ Biennial weeds are those weedy plants that live for two years. In the first year they attain their full vegetative growth. They produce flowers and set seeds in the second year, after which they die off. In comparison to annual and perennial weeds, the number of biennial weeds is much limited.

☆ Common examples are- *Cichorium intybus, Cirsium vulgare, Daucus carota, Gnaphalium obtusifolium,* and *Senecio jacobaea.* A weed like *Malva neglecta* exhibits features of both annual and biennial weeds. Also, sometimes *Cichorium arvense* can bolt every year to behave as an annual species. In crop fields the control measures of biennial weeds are planned along with those of the annual weeds.

☆ Perennial weeds persist for more than two years; more usually for a number of years. Their aerial parts may wither every year at the end of a season after producing flowers and seeds, but new shoots develop again from the underground vegetative organs like roots, rhizomes, tubers, stolons, and bulbs at appropriate time.

☆ When major portion of the underground vegetative organs of perennial weeds is limited to the plow layer of soil, these are sub-classed as shallow weeds, *e.g. Cynodon dactylon, Agropyron repens, Scirpus* sp., *Cyperus esculentus, Allium vineale*, and *Rumex crispus.* Many perennial weeds have their underground system located up to 1 m depth or deeper.

☆ These are sub-classed as deep weeds, *e.g. Cirsium arvense, Convolvulus arvensis, Sonchus aroensis, Solarium elaegnifolium, Euphorbia esula*, *Mikania micrantha, Pluchea lanceolata, Cyperus rotundus,* and Sorghum halepense. The perennial weeds are very difficult to control. Even the best translocated herbicide available today cannot reach all their underground parts.

☆ Tillage may temporarily destroy the aerial shoots of perennial weeds to give a clean look to the land, but actually it spreads their underground parts to the fresh spots by fragmentation. Each piece of rhizome, tuber, or bulb of perennial weed then grows into a new plant which finally takes the shape of a full-fledged infestation.

☆ It may be noted here that when a new plant of a perennial weed begins from its seeds, at that stage it can be treated for its control like an annual weed. The same is true for some perennial weeds like *Oxalis striata, Rumex crispus,* and *Plantago lanceolata* which reproduce only by seeds.

2. Grasses, Sedges, and Broadleaf Weeds

☆ This is perhaps the oldest and yet the most common classification of weeds used by the weed scientists. It took its roots from the time when the first successful herbicide 2,4-D (and MCPA) was found to easily kill the broadleaf weeds in cereals like wheat, barley, and oat, without damage to the crop plants. Despite discovery of numerous new herbicides since then, this classification of weeds is still in vogue because most herbicides are selective to either of the two categories of weeds, *viz.*, the grasses and the broadleaf's.

☆ It may be brought home here that in the weed science literature sometimes grasses are referred to as monocot weeds, and the broadleaf weeds as dicot weeds. Strictly speaking, this nomenclature is not very correct since the broadleaf weeds like *Eichhornia crasipes, Commelina benghalensis, Cyanotis axillaris,* and *Monochoria hactata,* are all monocot plants.

☆ Likewise, sometimes one tries to talk of grasses as narrow leaf weeds. This is also not always correct since sedges (*Cyperus* spp.) and cattlails (*Typha* spp.), are also narrow leaf plants although they don't belong to the family of grasses (Poaceae). Yet, for all practical purposes, the designation of weeds as broadleaf's, grasses, and sedges continues with the weed control scientists, worldwide.

3. Woody and Herbaceous Weeds

☆ The woody weeds, as the name suggests, have woody or semi- woody, rough stems. They are also called brush weeds. These are broadleaf, largely perennial shrubs and under-shrubs. They are problematic in grasslands, forests, uncropped areas, and fallow fields, although some of these like *Zizyphus* spp. can settle in crop fields too.

☆ Woody weeds had to be identified as a separate class of weeds because the two popular phenoxy herbicides, 2,4-D and MCPA, destroyed only the herbaceous broadleaf weeds but not the woody ones. Some common brush weeds of importance in agriculture are: Lantana (*Lantana camara*), mesquite (*Prosopis juliflora*), wild Indian plum (*Zizyphus rotundifolia*), poisonoak (*Rhus* spp.), blackberry (*Rubus allegheniensis*), and multiflora rose (*Rosa multiflora*). It may be noted that many brush weeds reproduce only by seeds, despite being perennials.

☆ In variance with the woody weeds, herbaceous weeds have green succulent stems, and are of common occurrence on farmlands. These weeds are easier to control than the woody weeds. Sometimes, semi-woody weeds are considered as different from the woody weeds. Camelthorn (*Alhagi pseudalhagi*) and arrowood (*Pluchea lanceolata*) are two such semi-woody, troublesome weeds with us today.

4. Parasitic Weeds

☆ There are certain plants which parasitise, fully or partially, on specific crop plants, for example, dodder (*Cuscuta* spp.) on lucerne, broomrape (*Orobanche* spp.) on tobacco, and witch weed (*Striga* spp.) on sorghum and pearlmillet. Such weeds are called parasitic weeds. They attach themselves either to the roots or to the shoots of the host plants and survive on food material available in them.

☆ The parasitic weeds are host-specific; they cannot survive in the absence of their host plants. Many weeds are also found to act as host plants to particular parasitic weeds, thus making it possible for them to survive even outside the crop fields.

5. Crop-Associated and Crop-Bound Weeds

Crop-associated weeds show their preference for association with certain crops for several reasons other than parasitism, as follows:

(i) Adaptation to Specific Habitat

Weeds like chicory (*Cichorium intybus*) and swinecress (*Coronopus didymus*) usually infest crops like lucerne and potato which offer moist, cool, and shady habitat suited to such weeds.

Mimicry-Some weeds survive with particular crops because of similarity in their foliage with the crop plants during their vegetative growth period. Canarygrass (*Phalaris minor*) and wildoat (*Avena fatua*) are good examples of mimics in wheat and barley where they easily escape the hoe of the farmer. Likewise, wild rice (*Oryza sativa* var. fatua), *Echinochloa colonum* and *E. crusgalli* survive well in cultivated rice by mimicry.

(ii) Easy Seed Contamination

Many weeds mature at the sometime and height as the crop they infest and are thus harvested and processed with it. This is perhaps the most common way our small grains are contaminated with many weed seeds. Crop-bound weeds, as different from the crop-associated weeds, are parasitic in nature. These complete their life cycles, fully or partially, on specific crop species only. Their control measures are developed keeping in view the nature of parasitism these depict.

When any alien weed becomes so aggressive as to fast displace the indigenous flora of the new ecosystem, it comes to be known as Invasive Alied Weed (IAW). In other words, an IAW threatens the native floral biodiversity of the area it infests. In India the major invasive alien weeds include: *Ageratum houstorium, Alternantliera philoxeroides, Chromolaena odorata, Eichhornia crassies, Lantana camara, Mikania micrantha, Mimosa* spp., *Cuscuta* spp., *Ipomoea carnea*, and *Phalaris minor*. The distinction between the native and alien weeds is of main interest to the biological weed control workers. They find the biological control of alien weeds comparatively easy by introducing their bioagents from the places of their origin. The quarantine people also need to become familiar with the alien weeds to be able to check their apparent entry in the new environment. The origin of many an alien weed is not yet known.

7. Facultative and Obligate Weeds

☆ Facultative weeds are those weeds which grow primarily in undisturbed or closed communities. But these may sometimes escape to the cultivated fields, for example, *Opuntia* spp. and *Parthenium hysterophorus*. Obligate weeds, on the contrary, occur primarily in the cultivated fields where the land is disturbed frequently.

☆ The obligate weeds cannot withstand competition from volunteer vegetation in a closed community of facultative weeds. If a cultivated land is abondoned permanently, sooner or later the obligate weeds give way to facultative weeds. For example- even a heavy infestation of a hardy, obligate weed like bindweed (*Convolvulus arvensis*) can be outgrown to the point of elimination if the field was abondoned sufficiently long (which, of course, is not possible on farm lands).

8. Noxious Weeds

☆ A noxious weed is a plant arbitrarily defined as being especially undesirable, troublesome, and difficult to control. The status of a plant as noxious weed will vary with the legal interpretation of a country or a state, as well as with the development of new weed control technologies.

☆ The noxious weeds have immense capacity to reproduce and disperse, and they adopt tricky ways to defy man's efforts to get rid of them. The noxious weeds are sometimes also referred to as special weeds and obnoxious weeds. Some common noxious weeds in India today are- *Cyperus rotundus, Cynodon dactylon, Parthenium hysterophorus, Eichhornia crassipes, Solanum elaegnifolium* and *Orobanche* spp.

☆ The status of a noxious weed changes with time. For instance, Striga is no more a noxious weed in the USA where it has been controlled effectively with ethylene. Similarly, *Salvinia molesta, Phalaris minor*, and *Cuscuta arvenis* are now not noxious in India since their control measures have been found.

☆ More recently, in Tamil Nadu wild brinjal (*Solanum elaegnifolium*), a noxious weed, has forced the farmers to replace their annual crops by the perennial plantations of coconut, eucalyptus, and casuarina where more rigorous weed control measures can keep this weed in control. *S. elaegnifolium* is a deep rooted, thorny, perennial broadleaf weed.

☆ In advanced countries there are strict laws to prevent the entry of new noxious weed seeds from outside the country, as well as the inter-state movement of the noxious weeds already present in some part of the country.

9. Objectionable Weeds

☆ Weeds which produce seeds that are difficult to separate once mixed with crop seeds are called objectionable weeds. Wild oat in wheat, dodder in lucerne, and *Polygonum plebejum* in cumin are few examples of such weeds.

☆ Special efforts are needed to prevent seeds of objectionable weeds from entering the crop harvests.

10. Aquatic Weeds

☆ Aquatic weeds grow within and around the fresh water bodies. They possess special features to withstand their partial or complete submergence in water. Aquatic weeds infest canals, ponds, lakes, irrigation channels, drainage systems, and the low level paddy fields.

☆ Some aquatic weeds of tropics are water hyacinth (*Eichhornia crassipes*), waterfern (Salvinia molesta), hydrilla (*Hydrilla verticillata*), and cattails (*Typha* spp.).

11. Industrial Weeds

☆ Weeds invading areas around buildings, highways, air strips, railway tracks, utility rights-of-way, fence-rows, industrial pipelines, electric and telegraph pole bases, observatory structures, gravel walks, petroleum farms, and several other like situations are called Industrial weeds. These form part of non-crop land weeds.

☆ These are adapted to undisturbed soil environment and various biotic stresses. The industrial weeds prevent proper use of highway and airstrips, involve fire hazards, hide pipelines, and hinder approaches to poles and buildings. Usually, the long term control and eradication programmes initiated against industrial weeds, involve non-selective herbicides and soil sterilants.

12. Grassland Weeds

☆ As the name indicates, weeds belonging to this class invade grasslands, rangelands, and permanent pastures, which offer a completely different ecological environment than the crop lands. The major difference between the two situations, from the stand point of view of weeds, is that while crop lands are frequently tilled and disturbed, the grasslands remain undisturbed for long period.

☆ The grassland weed species, however, must withstand frequent grazing and cutting, as well as trampling by the animals. Some grassland weeds are equipped with mechanisms to keep the animals away, like bitter leaves, poisonous foliage, prickly shoots, and hard stems.

☆ Although the scientists have classified weeds based on their ecological preferences and some other features, yet there are no hard and fast physical boundaries limiting their spread. Frequently, a weed considered to be typical of one agro-ecological situation can invade also the other, widely differing situation, depending upon its plasticity.

☆ Still, the classification of weeds is essential and useful in planning their management strategies. For instance, the industrial weeds will require for their destruction the use of non-selective herbicides, as against the crop land weeds which are treated with highly selective herbicides. In grasslands on the other hand, partially selective herbicides can also be used. Adoption of proper crop rotations offers the best way to avoid the crop-associated and crop-bound weeds in agriculture.

13. Alien and Invasive Alien Weeds

☆ When a weed is allowed to move from the place of its origin to a new region, and it establishes itself there, it becomes an alien weed in its new environment. In India, there is a large number of introduced weeds with us which were brought chiefly from tropical America and Australia, during the eighteenth and nineteenth centuries. By now all these introduced weeds have become so widespread in the country that they look like indigenous species.

☆ Several factors have led to the introduction of alien weeds in India. The major amongst these are- greatly increased travels across the countries, introduction of new crops, development of livestock industry, import of food grains and agricultural seeds, and the fast transport systems. In certain cases long distance travel by birds has also introduced new weeds. The seeds of carrotgrass (*Parthenium hysterophorus*), corncockle (*Agrostemma githago*), and *Solatium elaegnifolium* were believed introduced in India with food grains imported from the USA.

☆ Viable seeds of alligator weed (*Alternanthera philexeroides*) were brought accidentally with packing material from South America during the Second World War. *Chrotnolaena odorata* is native to America from where it was navigated to India. Waterhyacinth (*Eichhornia* crassipes) and salvinia (*Salvinia molesta*) were brought into India as ornamental plants. Fruits and seeds of *Clidemia hirta*, a tropical American species first noticed in India in Kerala rice fields, were introduced perhaps by the bird Indian myna.

☆ Lantana (Lantana camara) seeds were also spread from Sri Lanka by this bird. Cocklebur (*Xanthium strumarium*) is a native of America. Its inflorescence sticks to the body of animals and, perhaps, it was introduced that way in India with the animal imports. Puncturevine (*Tribulus terrestris*) was brought to India from the Mediterranean region with aircraft tyres.

Weed Ecology

☆ **Ecology** is defined as the study of how organisms interact with one another and their physical surroundings. **Weed-crop ecology** is the study of how weeds interact with a crop, humans, and their physical surroundings.

☆ Weeds typically invade bare, open areas where more desirable plants tend not to grow. Annual grass weeds typically invade first followed. Next annual broadleaf weeds typically invade. Those are typically followed by perennial grass weeds, perennial broadleaf weeds, and brambles and vines. Trees are typically the last species to invade an area.

☆ Growth habits and seeding are characteristics of weeds that most affect their ability to survive and thrive.

Growth Habits of Weeds

☆ Weeds are typically adaptive, resilient species of plants. They can live in a variety of habitats. They adapt to changing environments, thrive under extreme conditions, establish populations in strange places, and succeed in disturbed environments where other plants cannot.

☆ Some weeds have specialized, modified plant parts to help them thrive where cultivated plants cannot. Some have shallow roots to help them grow in compacted soil. Some have prostrate growth patterns to avoid being damaged by mowing. Others that need to survive drought conditions have reduced leaf area and aerial parts.

Seeding of Weeds

☆ Like other weed growth habits, the seeding process for weeds has also adapted itself for survival. Weed can produce a lot of seeds in a single growing season. They can produce thousands, sometimes tens of thousands, of seeds during a growing season.

☆ Many have evolved such that their seeds can remain buried and dormant for years until conditions are favorable for germination. The seeds for some plants can remain viable for 40, 50, or 70 years. The seeds of many weeds can continue to mature days after the parent plant has been pulled from the ground.

☆ Weed seeds can be disseminated in a variety of ways. Wind, water, and animals such as birds or humans help to spread their seeds. Some weed seeds have tufts of hair or wing-like structures enabling them to be carried by the wind. Other are lightweight and/or have an oily film allowing them to be carried by rainwater runoff. Still others have physical characteristics like burs, hooks, and other physical properties making it easy for them to attach themselves to animals and humans for transport.

☆ Weed seeds are often mixed in with crop seeds since they are sometimes hard to see or separate from crop seeds making it important to use weed-free seeds for planting in a garden. Cultivation of the land can also lead to the spread of weeds.

Crop-Weed Competition

☆ Competition is struggle between two organisms for a limited resource that is essential for growth. Water, nutrient, light and space are the major factors for which usually competition occurs.

☆ Competition between crop plants and weeds is most severe when they have similar vegetative habit and common demand for available growth factors.

☆ Weeds appear much more adapted to agro-ecosystems than our crop plants. Without interference by man, weeds would easily wipe out the crop plants.

☆ Generally, an increase in on kilogram of weed growth will decrease one kilogram of crop growth.

Principles of Crop Weed Competition

☆ ***Competition for nutrients:*** It is an important aspect of crop weed competition. Weeds usually absorb mineral nutrients faster than crop plants. Usually weeds accumulate relatively larger amounts of nutrients than crop plant Nutrient removal by weeds leads to huge loss of nutrients in each crop season, which is often twice that of crop plants. Amaranthus accumulate over 3 per cent nitrogen in their dry matter and this fall under category of nitrophylls. *Digetaria* spp. accumulates more phosphrus content of over 3.36 per cent. Chenopodium and Portuluca are potassium lovers, with over 4.0 per cent K_2O in their dry matter. *Setaria lutescens* accumulates as high as 585 ppm of zinc in its dry matter. This is about three times more than by cereal crop.

☆ ***Competition for moisture:*** Crop weed competition becomes critical with increasing soil moisture stress. In general for producing equal amount of dry matter weeds transpire more water than field crops. Therefore, the actual evapotranspiration from the weedy crop fields is much more than the evapotranspiration from a weed free crop field. Consumptive use of *Chenopodium album* is 550mm as against 479mm for wheat crop. Further it was noted that weeds remove moisture evenly from up to 90 cm soil depth. While the major uptake of moisture by wheat was limited to top

15 cm of soil depth. Weeds growing in fallow land are found to consume as much as 70- 120 ha mm of soil moisture and this moisture is capable of producing 15 -20 q of grain per ha in the following season.

☆ *Competition for light (Solar energy:)* Plant height and vertical leaf area distribution are the important elements of crop weed competition. When moisture and nutrients in soil are plentiful, weeds have an edge over crop plants and grow taller. Competition for light occurs during early crop growth season if a dense weed growth smothers the crop seedlings. Crop plants suffer badly due to shading effect of weeds. Cotton, potato several vegetables and sugarcane are subjected to heavy weed growth during seedling stage. Unlike competition for nutrients and moisture once weeds shade a crop plant, increased light intensity cannot benefit it.

☆ **Competition for space (CO_2):** Crop-weed competition for space is the requirement for CO_2 and the competition may occur under extremely crowded plant community condition. A more efficient utilization of CO_2 by C_4 type weeds may contribute to their rapid growth over C_3 type of crops.

Critical Period of Crop-Weed Competition

The period at which maximum crop weed competition occurs called critical period. It is the shortest time span in the ontogeny of crop when weeding results in highest economic returns.

Factors Affecting Weed-Crop Interference or Critical Period of Crop Weed Competition

1. Period of weed growth.
2. Weeds/crop density
3. Plant species effects:
   a) Weed species
   b) Crop species and Varieties.
4. Soil and climatic influence.
   a) Soil fertility
   b) Soil moisture status
   c) Soil reaction
   d) Climatic influences.
5. Cropping practices.
   a) Time and method of planting crops
   b) Method of planting of Crops
   c) Crop density and rectangularity.

Period of Weed Growth

Weeds interfere with crops at anytime they are present in the crop. Thus weeds that germinate along with crops are more competitive. Sugarcane takes about one month to complete its germination phase while weeds require very less time to complete its germination. By that time crop plants are usually smothered by the weeds completely. First 1/4 - 1/3 of the growing period of many crops is critical period. In direct sown rice more severe weed competition than transplanted rice. However in a situation, where weeds germinate late, as in dry land wheat and sorghum, the late stage weeding is more useful than their early weeding.

In general for most of the annual crops first 20-30 days weed free period is very important.

Weeds/Crop Density

Increasing weed density decreasing the crop yields. The relation ship between the yield and weed competition is sigmoidal. In rice density of Joint vetch (*Aschynomene virginica*) and barnyard grass, if it is > 10 plants/m^2 rice yields were reduced by 20 and 11 q/ha respectively.

Crop Density also Effect the Weed Biomass Production

Increase in plant population decreases weed growth and reduce competition until they are self competitive for soil moisture and other nutrients. In wheat reduced row spacing from 20 to 15 cm reduced the dry matter yield of *Lolium* and *Phalaris* spp. by 11.8 per cent and 18.3 per cent respectively.

Plant Species Effects Weed Species

Weeds differ in their ability to compete with crops at similar density levels. This is because of differences in their growth habits and to some extent due to allelopathic effects. At early stage of growth, cocklebur (*Xanthium strumarium*) and wild mustard (*Brassica* spp) are better competitor for crops than many grasses because of their fast growing leaves that shade the ground heavily. In dry areas perennial weeds like Canada thistle (*Cirsium arvense*) and field bind weed (*Convolvulus arvensis*) were more competitive than annual weeds because of their deep roots and early heavy shoot growth. Composite stand of weed sp is always more competitive than a solid stand of single weed spp.

Crop Species and Varietal Effects

They differ in their competing ability with weeds. Among winter grains the decreasing order of weed competing ability is barley > rye > wheat > oat. In Barley it may be due to more extensive root growth during the initial three weeks.

Fast canopy forming and tall crops are more competitive than slow growing short stature crops (sorghum, maize, soybean, cowpea).because of their slow initial growth. Late sown dwarf wheat is affected by the late germinating weeds like canada thistle and wild safflower. (*Carthamus oxycantha*) and *Phalaris minor* even though they escape an initial flush of weeds.

Varieties

Smothering crops grow very fast during early stages. Cowpea and horse gram are tolerant to weed competition. When we compare the crop-weed competition between two varieties of groundnut, in spreading groundnut (TMV-3) the yield loss is 15 per cent in weedy plots compared to bunch groundnut (TMV-2) where yield loss is 30 per cent. This is due to smothering effect of spreading groundnut. Like wise long duration rice is more competitive than short duration rice varieties. Wild oat growth increase with increase dwarfness of wheat plant.

Soil and Climatic Influence

a) Soil Fertility

Under limited nutrient conditions, competition exists between the crop and the weed. Soil type, soil fertility, soil moisture and soil reaction influences the crop weed competition. Elevated soil fertility usually stimulates weeds more than the crop, reducing thus crop yields. Method and time of application of fertilizers to crop determining whether added fertilizer will suppress or invigorate weed growth in fields.

Application of fertilizers during early crop growth season when weed growth is negligible was more beneficial. Band application of fertilizers to the crop will be inaccessible to inter row weeds.

b) Soil Moisture Status

Weeds differ in their response to available moisture in soil. Russian thistle Salsola kali showed similar growth in both dry soils and wet soils; where as large crab grass *Digitaria sanguinalis* produce more growth on wet soil. When fields are irrigated immediately after planting then weeds attain more competitive advantage over crops. If the weeds were already present at the time of irrigation, they would grow so luxuriantly as to completely over cover the crops. If the crop in irrigated after it has grown 15 cm or more in a weed free environment irrigation could hasten closing in of crop rows, thus suppressing weeds. In water logged soils weeds are more competitive than crop plants. In submerged conditions in rice, weeds are put to disadvantage to start with. But if there is a break in submergence, the weeds may germinate and grow more vigorously than the crop, even if fields were submerged later.

c) Soil Reaction

Abnormal soil reactions (very high or very low pH) often aggravate weed competition. Weeds offer intense competition to crops on abnormal pH soils than on normal pH soils. In acid soils *Rumex acetosella* and *Pteridium* spp., saline alkaline soils Taraxacum stricta, Agropyron repens are the dominant weeds.

d) Climatic Influences

Adverse weather conditions per se drought, floods and extreme of temperature intensify weed-crop interference since most of our crop varieties are highly susceptible to such climatic influences where as the weeds are tolerant to their stresses.

Cropping Practices

a) Time of planting crops If the time of planting of a crop is such that its germination coincides with the emergence of first flush of weeds, it leads to intense weed-crop interference. Usually longer the interval between emergence of crop and weeds, lesser will be the weedcrop interference.

b) Method of planting of crops may also affect the weed-crop competition. Weed seeds germinate most readily from top 1.25 cm of soil, though it is considered up to 2.5 cm depth. Avena, barnyard grass, *Xanthium* and *Vicia* spp. may germinate even from 15 cm depth. Therefore planting method that dries up the top 3-5 cm of soil rapidly to deny weed seeds opportunity to absorb moisture for their germination and usually post pone weed emergence until first irrigation. By that time crop establishes well and competes with weeds.

Critical Period of Crop-Weed Competition and Yield Reduction in some Crops

Crop	Critical Period of Crop Weed Competition	References
Rice	15-45 DAS	Reddy and Reddy
Wheat	30-50 DAS	Choudhary (2008)
Maize	15-35DAS	Reddy and Reddy
Sorghum	15-45 DAS	Bharti (2009)
Soybean	14-45 DAS	Surianto et al. (2017)
Groundnut	15-35 DAS	Reddy and Reddy
Sugarcane	2-120DAS	Akanksha (2010)
Pigeon pea	30-60 DAS	Ali M. (2017)
Geen gram	20-40 DAS	Sheoran et al. (2008)
Black gram	30-45 DAS	Vivek et al. (2008)
Chickpea	15-60 DAS	Singh and Singh, (1992)
Field Pea	20-60 DAS	Singh et al. (2016)
Lentil	20-60 DAS	Jamin et al. (2012)

Weed Control

☆ Most gardens and landscapes require a weed management program to be followed to control weeds and prevent them from competing with cultivated plants. The eradication, or complete elimination, of weeds is virtually impossible, so the goal of such programs is typically to control or minimize the occurrence of weeds.

☆ Most weeds can be controlled with integrated weed management. The four key components of such a program are prevention, cultural control, mechanical control, and chemical control.

Identifying Weeds

☆ The first step in any weed managements system should be to identify weed species that are or might become problematic. Pictures, vegetative characteristics, and line drawings in books and online resources are typically used to identify weed species. However, care must be taken when doing so because weeds may look very different during each of their stages of development, during different seasons, under different conditions, and in different settings in which they grow.

☆ To identify weeds, you should first look for unique characteristics to help narrow the list of potential weeds. This could include a variety of things like the conditions under which they are growing (wet vs. dry), whether its growth is erect or prostrate, leaf arrangements (alternate, opposite, whorled, simple vs. compound, *etc.*), broad vs. narrow leaves, smooth vs. dentate margins, and more. There are many books and online resources available to assist you with identifying weeds.

Weed Prevention

☆ There are many things that can be done to prevent the spread of weeds. Not only do you want to prevent them from spreading within your own landscape, but if possible, between your site and others.

☆ Ensure that all shovels, mowers, and other equipment used are cleaned before using them at another site to prevent the spread of weed seeds. Lawn care companies should clean their equipment between clients.

☆ To the extent possible, you should maintain not only your landscape, but the areas around it. If your lawn, garden, or other landscape area is bordered by weed infested empty lots, ditches, or adjacent lawns, they can continue to re-infest your landscape. Try to keep these as free of weeds as possible by using nonselective herbicides. Convince neighbors that all should have an integrated weed management program to eliminate cross infestation.

☆ Mulching is a great annual weed deterrent for gardening beds. A 2- to 3-inch layer of mulch will prevent weed seeds from receiving the sunlight needed for germination. Mulch has the added benefits of keeping the soil beneath it moist, cool in summer, and warm in winter as well as beautifying your landscape. Be sure the mulch you use has been treated so that it is free of weeds.

☆ If you use compost or manure, be sure proper composting procedures are followed. The compost piles that reach 160° F will be free of most weeds. If you plant seeds, be sure that the seed packaging label states they are weed-free (or mostly weed-free). If transplanting trees or shrubs with burlap wrapped balls or container plants, make sure they are weed-free.

Cultural Methods of Controlling Weeds

☆ When selecting and planting in your garden and landscape, it is important to ensure that your plants have the best chance to grow and thrive. Many things go into this such as picking the right type of plants for the conditions and growing zone, selecting disease resistant cultivars, planting during the correct time of year, and ensuring the soil is properly maintained with the correct pH, proper aeration, fertilization to name a few. When desirable plants grow and thrive in your garden and landscape, weeds have less chance of spreading and taking hold.

☆ Spacing your plants so that roughly 80 per cent of the soil is shaded can drastically limit most weed growth. Plant shrubs and other flowers densely so little soil is exposed to the sunlight needed for weeds to germinate. Care should be taken not to plant too densely, as this could lead to poor air circulation, diseases, and insect infestations. Crops should be planted in wide rows side by side to shade the soil between.

☆ Moisture can cause desirable plants to fail while providing an environment for weeds to thrive. So, it is important that landscapes and crops are not overwatered. Allow the gardens, landscapes, and lawns to dry thoroughly between each watering. Plants should be grouped based on their watering requirements so that when giving one the required amount of water, you are not overwatering others. The proper used of irrigation heads and driplines can ensure different zones of a garden or landscape each get the correct amount of water.

Mechanical Methods of Controlling Weeds

☆ Properly mowing your lawn can promote grass heath and control weeds. Never cut more than 1/3 of the grass height in a single mowing. Ensure that mower blades are sharp so that the cuts to the grass are clean and do not injure the grass blades. Injuries to the grass make it more susceptible to be overtaken by weeds, insects, and disease.

☆ Hand pulling weeds can be one effective method for controlling annual weeds in smaller areas. It is less effective for perennial weeds because reproductive parts of the weed could break off and t behind in the soil. When pulling weeds, care should be taken to minimize soil disturbance which can lead to the germination of more seeds. Pulling weeds should be done when the soil is moist such as after a light watering.

☆ As mentioned previously, mulching is an effective way of preventing the emergence of annual weeds. It is easy to implement as a mechanical control and relatively inexpensive. Mulch prevents weeds from germinating and growing by excluding sunlight.

☆ Tilling or turning the soil is often used to prepare the soil for planting and to mix in organic material. It is also effective for controlling weeds. A hand operated tiller or tractor pulled plow can be used to till or turn the soil, depending on the size of the job. Doing so damages the vegetative parts of the soil and its roots. The roots become exposed and dry out. This is very effective with younger plants, but repeated tilling may be required for more established perennial weeds. By repeated tilling the soil, even mature perennial weeds will eventually deplete their food store. Even though turning the soil may lead to the germination of additional weeds whose seeds were previously dormant, repeated tilling will eventually kill those plants and deplete the seed bank as well.

☆ Soil solarization is accomplished by covering the soil with clear or black plastic. While clear plastic typically produces higher soil temperatures, black plastic excludes light from reaching weed seeds and seedlings, preventing germination and photosynthesis. It should be implemented when the soil is most and during the summer months. Soil solarization traps heat under the plastic which kills weed seeds as well as nematodes and other disease-causing organisms. It is most effective against cool-season weeds.

☆ Fire can be used as a method of controlling weeds. Not only can it kill the dry mature plant matter and new growth, but it can often kill buried weed seeds as well. Burning may be less effective if the buried weed seeds are dry. It is often used to control weeds in ditches and along the sides of roads. Flaming is a method that uses a fan shaped blow torch to burn weeds in smaller areas like along fences. It is most effective on young weeds, but repeated flaming can also be effective against tough perennials.

Biological Methods of Weed Control

☆ Biological weed control is the utilization of living organisms to manage weeds. These living organisms can include arthropods like insects and mites, pathogens such as viruses, bacteria, fungi, and nematodes, and predators like birds, fish, and other animals. However, these methods are not typically used by amateur home gardeners and master gardeners.

☆ The method takes advantage of basic ecological interactions between organisms, such as predation, parasitism, pathogenicity and competition. Today, biological control is used primarily for controlling pests in crop cultivation.

☆ Advantages of biological control are that no artificial substances are added, and that pathogens/animals that develop resistance against biological control agents are rare. Biological control is an important component of integrated pest management

Chemical Methods of Weed Control

☆ In chemical weed control, chemicals called herbicides are used to kill certain plants or inhibit their growth. Chemical weed control is an option in integrated weed management that refers to the integrated use of cultural, manual, mechanical and/or chemical control methods. Most herbicides work by interacting and interfering with the metabolic functions and biochemical pathways of weeds causing irreversible damage to the weeds' tissue which eventually leads to its death.

☆ Many factors should be considered when selecting the best herbicide for a specific situation including the desirable plants or crops being grown, the weeds being targeted, how close the weeds are to desirable plants or crops, the growth stage of both the targeted weeds and desirable plants and crops grown nearby, the type of soil where the herbicide is being applied, the season when it is to be applied, the cost, how it will be applied, and how it might affect the surrounding environment.

Classification of Herbicides

Herbicides can be classified a variety of ways. Some common ways that they are classified include their mode of action, the site of action and the time of application.

There are at least 450 families of flowering plants and well over 350,000 different species. Only about 3,000 of them have been used by humans for food.

Fewer than 300 species have been domesticated, and of these, there are about 20 that stand between humans and starvation.

There are at least 100 species of great regional or local importance, but only a few major species dominate the human food supply.

Only about 15 plants provide most of the food that humans have consumed for many generations.

Twelve plant families include 68 per cent of the 200 species that are the most important world weeds (Holm, 1978)

Herbicide Classifications by Mode of Action

☆ There are two major categories of herbicides when classified based on their mode of action: contact herbicides and systemic herbicides. Contact herbicides affect the portion of the plant with which they make contact. They are not translocated to other parts of the plant, so it is important that they be applied in sufficient quantity to thoroughly cover the foliage.

☆ Systemic herbicides are translocated to other growing parts of the plant via its vascular system after entering the plant through its foliage or roots. Most systemic herbicides are applied to foliage rather than soil.

☆ Herbicides are also classified based on their selectivity. Selective herbicides control specific species of weeds or category of weeds as specified on the label, without harming or affecting other surrounding plants significantly.

☆ For example, a broadleaf weed killer might target the broadleaf category of weeds while not damaging the surrounding turf.

☆ Nonselective herbicides generally kill any plant with which they come into contact. Glyphosate or Roundup is an example of a non-selective herbicide.

Herbicide Classifications by Application Timing

☆ Herbicides can be classified based on when they are applied. There are four such classifications: pre-plant incorporated or farrow application, pre-plant application, pre-emergence, and post emergence. Some of these terms normally refer to crop herbicides but may also apply to landscaping and gardening use.

☆ Pre-plant incorporated herbicides or farrow application herbicides must be plowed or tilled into the soil before planting to be effective. If left on the soil surface, these highly volatile herbicides will quickly evaporate into the air. This is usually applied in an agricultural setting well in advance of sowing where crops such as tomatoes, soybeans, corn, and strawberries are being grown commercially.

☆ Pre-plant herbicides are applied prior to sowing but do not require incorporation into the soil. They are typically applied a few days before sowing.

☆ Pre-emergent herbicides are applied after a crop, garden, or turf is planted or is established which targets weed seeds before they can germinate. These herbicides kill weeds either by preventing the weed seeds from germinating or by killing the seedling before it emerges from the soil. These are generally selective herbicides and must be applied at the correct time since they are typically only effect for 10-12 weeks. To be effective, the application area should be watered to move the chemical into the soil. Otherwise, results will be poor.

☆ Post emergent herbicides are applied after weeds have emerged from the soil. They are applied directly to the foliage of existing weeds and therefore do not provide residual protection against the emergence of new weeds. Apply to the point where the herbicide begins to run off the foliage and stop. Any liquid that drips off the plant's foliage is wasted.

✫ The application of post emergent herbicides is most effective when the plant is actively growing rather than when it is seeding or undergoing stress from mowing, heat, drought, or cold weather. After application, you should generally let the herbicide dry on the foliage for at least 6 hours before exposing to rain or irrigation as early watering can wash the herbicide from the plant, though there are some post emergent herbicides that are rain fast and dry within an hour or so. These herbicides tend to be most effective when applied during times when the temperature is between 60° and 85° F. Temperatures colder than 60° can slow the translocation of the herbicide within the plant while temperatures above 85° F can cause harm to desirable plants not being targeted by the herbicide.

Herbicide Formulations

✫ Herbicides can be purchased in many forms. They are formulated to make handling, mixing, and applying them easier. A letter designation can usually be found on the label indicating the formulation.

✫ The major formulation categories and corresponding letter designation are: soluble concentrates (SC), soluble powders (SP), emulsifiable concentrates (E or EC), wettable powder (W or WP), dry flowables (F or DF) and water dispersible granules (DG or WDG), granules (G) and pellets (p), and ready to use (RTU).

✫ Soluble concentrates mix readily with water requiring no agitation. They stay in solution and do not separate or settle in their containers once mixed. They are non-volatile, non-abrasive, do not clog screens or nozzles, and the equipment cleans easily. Soluble powders are the same as soluble concentrates only in powder form making them easier to store, transport, and handle.

✫ Emulsifiable concentrates contain emulsifiers to form stable oil: water mixtures. These emulsifiers wrap around oil soluble chemicals to suspend them in a water-based or aqueous solution like the way small droplets of oil or fat are suspended in water to form milk. In fact, when emulsifiable concentrates are mixed with water, the resulting suspension is "milky white" in color.

✫ Wettable powder is the name attributed to the original dry herbicide formulations. They can be applied as is directly to plants or mixed in water. These herbicides are problematic due to how fine they are (the consistency of flour or talcum powder) which makes them dusty and easy to inhale. The powder tends to stick together and come out the bag in clumps when attempting to weigh and mix them creating clouds of dust.

✫ Dry flowables and water dispersible granules are terms used for a new and improved breed of wettable powder. There is really no difference between dry flowables and water dispersible granules formulations other than what their manufacturers call them. The powder is pre-clumped together into aggregates or granules which disperse when added to water.

☆ Granules and pellets, like most dry products, are attached to inert particles used as the carrier. These formulations have large particles with a low concentration of herbicide. They can be applied directly to the soil from their packaging using a spreader without any preparation or need to dissolve them in water.

☆ Ready to use herbicides are premixed in the carrier and can be applied directly from the purchased container. These containers often come with a spraying nozzle already attached.

Herbicide Safety

☆ Once you have identified the desirable plant and problem weed, then you should select an appropriate labeled herbicide for controlling the weed without harming the plant. Read and understand the entire label on the herbicide container as they always provide information on the plants to which it can be applied, sites where it can be used, and which weeds it will control.

☆ The packaging label is the best source of information on how to use the product correctly and safely. Follow all instructions on the label, never applying more than is recommended which increases the risk of injury to desirable plants.

Harmful Effects of Weeds

Weeds are harmful in many ways. The damages caused by them are as under:

☆ Reduction in crop yield: Weeds compete with crops for water, nutrients and light. Being hardy and vigorous in growth habit, they grow faster than crops and consume large amount of water and nutrients, thus causing heavy losses in yields

☆ Increase in the cost of cultivation: Tillage operations are done to control weeds and it is generally estimated that on an average about 30 percent of the total expenditure for crop production is on tillage operations and more labour is employed for weeding. This results in increasing cost of cultivation and reducing the margin of net profit.

☆ The quality of field produce is reduced: When the crop is harvested from a weedy field the seeds of weeds get mixed with the main crop which results in lowering the quality of the produce *e.g.* seeds of weeds in wheat, gram *etc.* Similarly, bundles of many leafy vegetables like methi or palak contain green plants of weeds. They fetch lower prices in the market.

☆ The quality of the livestock products is reduced. Certain weeds *e.g.* Hulhul when eaten by milch cattle impart an undesirable flavour to milk. Similarly weeds like gokhru get attached to the body of the sheep and impair the quality of wool. Certain poisonous weeds like Datura may cause death of cattle if they are unknowingly eaten by cattle.

☆ Weeds harbour insect, pests and diseases: Weeds either give shelter to various insects, pests and diseases or serve as alternate hosts.

☆ Weeds check the flow of water: Weeds block drainage and check the flow of water in irrigation channels and field channels thereby increasing the seepage losses as well as losses through overflowing. The irrigation efficiency is also reduced.

☆ Weed secretions are harmful: Heavy growth of certain weeds like quack grass or motha lower the germination and reduce the growth of many crop plants. This is said to be due to the presence of certain phytotoxins in these weeds.

☆ Weeds are harmful to human beings: certain weeds cause irritation of skin, allergy and poisoning in human beings.

☆ Weeds cause quicker wear and tear of farm implements; they get worn out early and cannot work efficiently unless they are properly sharpened or mended.

☆ Weeds reduce the value of the land: Agricultural lands which are heavily infested with perennial weeds like kans always fetch less price, because such lands cannot be brought under cultivation without incurring heavy expenditure on labour and machinery.

Beneficial Effects of Weeds

Weeds are beneficial in many ways. Some of the advantages of weeds are:

☆ Add organic matter to the soil when incorporate into the soil.

☆ Increase soil fertility when incorporated.

☆ Induce soil formation by rapid weathering.

☆ Improve soil structure spreading of weed roots change the soil structure and improve the physical condition of soil due to proper percolation water logging will be prevented.

☆ Serve as food (Bothua, Shaknotey, Amrul *etc.*), feed (Durba, Mutha, Chapra *etc.*) and medicine (Thankuni, Durba *etc.*).

☆ Serve as raw materials for public utilities as fuel, fencing and roofing materials.

☆ Help in controlling soil erosion.

☆ Serve as water purifier. *e. g.* Water hyacinth.

☆ Serve as a source of genetic materials.

✰ Some weeds can be used as indicators of air pollution. *e.g.* Wild mustard is an extremely sensitive indicator of NH_3, NO_2 present in air.

✰ Help in soil reclamation. *e.g.* Durba and Shialkata, when incorporate into the soil may reclaim alkaline soil.

✰ Valued as religious and ritual purposes.

✰ Valued as ornamental plant.

✰ Act as a host for predatory insect.

✰ Leguminous weed can be used as green manure before they set seeds.

✰ Some weeds fix atmospheric nitrogen in paddy soil. *e.g.* BGA

Chapter 11

Herbicides

Introduction

☆ Herbicides are chemicals used to manipulate or control undesirable vegetation. Herbicide application occurs most frequently in row-crop farming, where they are applied before or during planting to maximize crop productivity by minimizing other vegetation. They also may be applied to crops in the fall, to improve harvesting.

☆ Herbicides are used in forest management to prepare logged areas for replanting. The total applied volume and area covered is greater but the frequency of application is much less than for farming (Shepard *et al.*, 2004).

☆ In suburban and urban areas, herbicides are applied to

Herbicides or weedkillers belong to a class of pesticides that are used in the management of undesired plants in the areas of agriculture, landscaping, forestry, gardening, and industry. The control of weeds and other unwanted plants in a cost-effective manner is very important to agriculture and other related industries. Herbicide use, though essential for limiting and eliminating the weed populations, poses its own set of problems and risks; its use must be minimized to account for the desired economic and environmental effects. The problem associated with weed control is amplified due to herbicide resistance that some of the weed species have developed over the course of time due to the overuse of herbicides or their evolution (process of natural selection) toward favorable conditions.

lawns, parks, golf courses and other areas. Herbicides are applied to water bodies to control aquatic weeds. These weeds can impede irrigation withdrawals or interfere with recreational and industrial uses of water (Folmar *et al.,* 1979).

☆ The potential effects of herbicides are strongly influenced by their toxic mode of action and their method of application. The molecular site of action is challenging to predict because structural associations have not been identified (Duke 1990), but modes of action are well-established.

☆ Herbicides can act by inhibiting cell division, photosynthesis or amino acid production or by mimicking natural plant growth hormones, causing deformities (Ross and Childs 1996). Application methods include spraying onto foliage, applying to soils and applying directly to aquatic systems.

History of Herbicides

☆ History of herbicides starts with the advent of agriculture. In early days of farming humans had to spent good amount of their energy to remove weeds from arable lands so that suitable conditions are provided for the optimum growth of desired crops.

☆ Simultaneously the thought about weed management or control started to occupy minds of the then farmers. Hay (1974) gave six stages in evolution of weed control:

❖ 10000 B.C. – Removing weeds by hand

❖ 6000 B.C. – The use of primitive hand tools to till the land and destroy weeds

❖ 1000 B.C. – Animal-powered implements like harrows

❖ 1920 A.D. – Mechanically-powered implements like cultivators, blades, harrows, finger-weeders, rotary-hoes, rod-weeders, *etc.*

❖ 1930 A.D. – Biological control, and

❖ 1947 A.D. – Chemical control, with the commercial development of organic herbicides such as 2, 4-D and MCPA.

Initially man used to remove weeds by hand. Around 6000 B.C. primitive hand tools replaced the use of bare hands to destroy weeds. Then it was the era of the use of animals like oxen and horses to pull harrows (tools) in 1000 B.C. The early Romans (164 B.C.) used salt to destroy the agriculture in Carthage. English farmers in sixteenth century used salt to selectively kill thistle in wheat fields and remove weeds from garden paths (Lowery 1987).

Use of chemicals as herbicide is not new. Chemicals, in crude form, such as oil wastes, rock salts, crushed arsenical ores, copper salts, and sulfuric acid have been in use for weed eradication from railway tracks, car roads and timber yards (Green *et al.,* 1987). These herbicides used to destroy all the plants and can be categorized as non-selective. Also, the treated plot of land remained toxic to the plants for good period of time. Unfortunately these chemicals could not be used in arable lands because of the adverse effect on crop plants. Thus, application of selective herbicide that specifically kill only weeds came into existence. Bolley in the United States, Schultz in Germany and Bonnett in France initiated research on way back in 1900 (Klingman *et al.,* 1982) and concluded that inorganic compounds and solutions of copper salts selectively control broadleaf weeds in cereals. When plant growth regulators were discovered nobody would have believed that these chemicals can be a probable herbicide. Plant growth was one the most important topics of study for the early Botanists. Darwin from England, Boysen-Jensen from Denmark did the famous experiment on oat coleoptile and phototropism. It was Went who found that the chemical present in oat coleoptile is an active growth regulator (Went and Thimann 1937). This gave an impetus to research in this area and now growth regulators as herbicides is one the key areas of research. Went once stated that, "When I worked 25 years ago with the growth hormone (2,4-D), I had many wild ideas about what it might do once it was available in large quantities, but I never dreamed that it would lead to the development of weed killers. This is an excellent example, how fundamental research may lead to the solution of very practical problems" (Andersen 1991). The chemical synthesis of 2, 4-dichlorophenoxyacetic acid (2, 4-D) was described by Pokorny in 1941. Thereafter other salts and esters of 2, 4-D was developed. This growth regulator came in prime news during World War II when scientists from United States and England initiated research on plant growth regulators. This actually started the 'chemical era' for the development of herbicides. Templeman and Sexton in 1940s reported that phenoxyacetic acids were toxic to dicotyledonous, but not monocotyledonous plants. Later, 2, 4-D came in commercial use in U.S.A. and MCPA for use in Europe from 1947 (Rao 2000). Farmers use 2, 4-D as a selective killer of broadleaf dicot plants but not monocots (grass species). DNOC (dinitro compound), a contact herbicide, was used in France during 1993 for selective weed control. It acted against annual weeds without damaging cereals appreciably. DNOC is not translocated in plants. It was not effective against perennial weeds because their root system remained unaffected, which could later develop shoots. DNOC and other dinitro compounds played a big part in increasing food production during World War II as herbicides (Cremlyn 1991). Success in the use of these herbicides persuaded the scientists and industries as well to put money and mind in herbicides related researches, which is still in progress.

Important Definitions

Selective Herbicides: The chemicals which kills or retards the growth of some plants with little or no injury to other plants.

Non-Selective Herbicides: These chemicals are toxic to all the plants or kill all kinds of vegetation.

Contact Herbicides: A herbicides which kills only those plants or retards the growth of those plants which comes in direct contact.

Translocated Herbicides: The herbicides which are absorbed by the one part of the plants and exert a toxic action to other parts. These are also known as systemic herbicides. These absorbed chemicals upset the plant growth and metabolic processes.

Soil Fumigants: They usually function as a vapour or gas that diffuse through the soil and have relatively short life in the soil.

Soil Sterilants: Any chemical which prevent the growth of green plants when present in the soil is considered as soil Sterilants.

Classification

Herbicides are classified/grouped in various ways *e.g.* according to the chemical family, activity, method of application, site of action or timing of application.

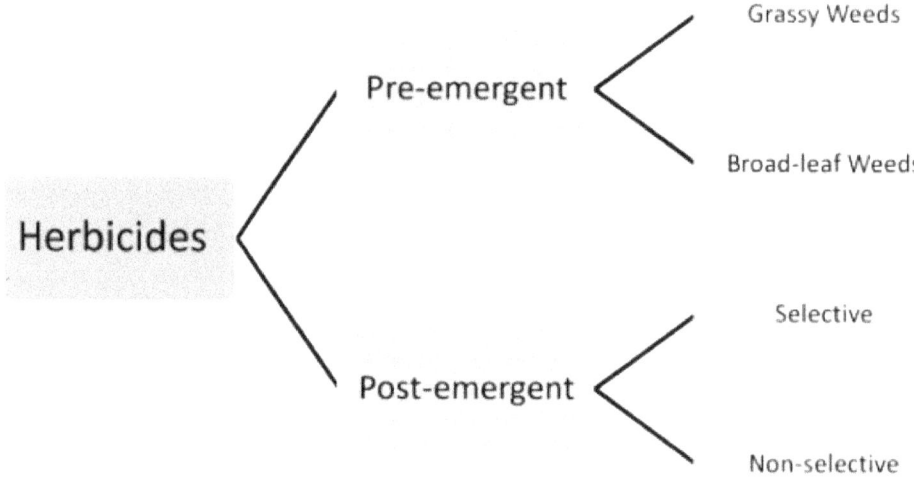

Herbicide Classification and Application.

Classification of Herbicides

(I) According to Selectivity of Herbicides

☆ Selective herbicides: 2, 4-D, Simazine, Atrazin, Butachlor, Pendimethalin, Fluchloralin *etc.*

☆ Non-selective herbicides: Diquat, Paraquat, Pendimethalin *etc.*

(II) Time of Application of Herbicides

⭐ Fallow application: Application of herbicides well in advance of sowing *i.e.* >10 days before sowing. It is applied for problematic weeds with higher dose.

⭐ Pre-plant application: Applied 2-4 days before sowing/planting *i.e.* Fluchloralin, Alachlor *etc.*

⭐ Pre-emergence: 1-4 days after sowing, *i.e.* Simazine, Atrazin, Butachlor, Pendimethalin *etc.*

⭐ Post-emergence: 30-40 DAS, *i.e.* 2, 4-D, Diquat, Paraquat, Isoproturon, Dalapan *etc.*

(III) Their Chemical Groups

Sl.No.	Chemical Groups	Associated Herbicides
1	Sulphonyl ureas	Sulpho sulfuron, Chlorimuron-ethyl, Meta sulfuron-ethyl
2	Aliphatic	TCA, Dalapan
3	Amide	Alachlor, Butachlor and Propanil
4	Bipyridiums	Paraquat, Diquat
5	Dinitroanilines	Fluchloralin, Pendimethalin
6	Chloro phenoxy compound	2, 4-D, 2,4,5-T etc
7	Triazines	Atazine, Simazine
8	Ureas	Monuron, Diuron
9	Dipheyl ether	Nitrophen, Oxyflorefen
10	Phenoxy phenoxy alkartoic acid	Clodinofop, Fenoxa prop-ethyl
11	Thiocarbamate	Benthiocarb
12	Organophosphorus	Glyphosate, Anilophos
13	Imidazolines	Imazethapyr, Imazapic

Application of Herbicides

(A) Soil Application

⭐ Soil surface application Most of the Triazines, urea's and amide group

⭐ Soil application Anilines group *i.e.* Fluchloralines

⭐ Sub-surface application → Only for deep rooted and perennial weeds

⭐ Band application Weeds in maize (spraying of Atrazine)

(B) Foliar Application

⭐ Blanket application→ Application of herbicides over the entire leaf area (only selective herbicides)

✰ Direct application→ Application of herbicides in between the crop rows directly towards weeds.

✰ Spot application→ Herbicide solution is poured on weeds in cropped and non cropped fields infested with abnoxious weeds in isolated patches.

✰ Basal application Brush wood and unwanted trees are treated with herbicides. Generally, the bark of the trees at the base of the stem up to 30 cm height is removed and a drenching spray of herbicides is given to the base.

Herbicidal Selectivity

✰ "When an herbicide is applied in a mixed plant population, herbicide harm or kills target weeds whereas crop-plants are not affected. This phenomenon is called selectivity"

✰ Selectivity is mainly depending upon weed nature and dosage of herbicide. Ex. Recommended dose of Atrazine (0.5-1.0 kg/ha) kills weeds of sorghum, means it acts as selective herbicide. But when Atrazine is applied at 10 kg/ha, it is non-selective in nature.

Trade Name of Different Herbicides

Chemical Name	Trade Name
Acifluorfen	Blazer
A crolein	Aqualin, Weedazol
Anilophos	Azalin
Atrazine	Atratof, Solaro
Alachlor	Lasso, Lazo
Butachlor	Machete, Delachlor
Benthiocarb	Saturn
Cholorimuron-ethy	Classic, Kloben
Chlosulfuron	Glean
Chlorimuron 10 per cent + Metasulfuron-methyl 10 per cent	Almix
Diuron	Cormex
Diquat	Reglone, Dextrone
Da lapan	Tafapan, Radapan
Ethoxy sulfuron	Sunrise
Fluchloralin	Basalin
Fenoxa prop-ethyl	Puma super, Whip super, Rice star
Glyphosate	Roundup
Linuron	Afalan
Metalachlor	Dual

Chemical Name	Trade Name
Metribuzine	Senor
Nitrofen	Toke E-25
Oxiflurofen	Goal
Oxadiazone	Ronstar
Paraquate	Gramoxone
Pendimethalin	Stomp
Propanil	Stam F-34
Simazine	Tafasine
Sulfosulfuron	Leader
2, 4-D	Plantgard
2, 4-DB	Butoxone

Formulation of Herbicides

☆ Soluble powder – 2, 4-D sodium salt, Dalapan, TCA

☆ Soluble Concentrates – 2, 4-D amine aster, Diquat, Paraquat

☆ Wettable Powder – Atrazine 80 per cent, Simazine 50 per cent, Isoproturon 70WP

☆ Liquid Suspension – Atrazine, Cyprazin, Nitralin

☆ Emulsifiable Concentration – 2, 4-D ester, Alachlor, Nitrofen

☆ Granules – Granules of Butachlor, 2, 4-DEE

Adjuvants

☆ The technical grades of herbicides are not applied in crop field directly. Most of them are formulated before use, by mixing the active ingredient(s) with inert(s) and/or other auxiliaries/adjuvants, to obtain a product which is effective, easy to handle and apply, possesses satisfactory shelf-life and is devoid of undesirable side effects (Parmar and Tomar, 2004). The auxiliaries or adjuvants present in a formulation are essential for the performance of most herbicides.

☆ ASTM (1998) has defined an adjuvant as 'a material added to a tank mix to aid or modify the action of an agrochemical, or the physical characteristics of the mixture' and in WSSA Herbicide Handbook (1994) it is defined as 'a substance in an herbicide formulation or added to the spray tank to modify herbicidal activity or application characteristics'. Thus, adjuvants enhance the performance of an herbicide in different ways, *viz.* wetting, spreading, deposit building, emulsifying, deflocculating the active ingredients, and many more. It may even enhance the bioefficacy of the active ingredients, thus it reduces the effective herbicide dose as much as 10-fold (Green and Green, 1993).

Different types of adjuvants used in herbicide formulations

Activator	*Acidifying Agent*	*Additive*
Antifoam/Defoam	Antifreeze	Attractant
Buffering agent	Binder	Coupler
Chelating agent	Compatibility agent	Colorant/Dye
Detergent	Deposition agent	Drift control agent
Dispersant	Emulsifier	Evaporation reducer
Foam marker	Humectant	Inert
Neutraliser	Modified seed oil	Preservative
Petroleum oil	Penetrator	Rainfast agent
Spreader sticker	Synergist	Safener
Surfactant	Translocation aid	Thickener
UV protectant	Vegetable oil	Wetting agent

Source: Green (2000).

☆ Based on the use there are different types of adjuvants (Table), among which surfactants are key adjuvants used in most of the formulations. Surfactant is a formulant which reduces the interfacial tension of two boundary surfaces, thereby increasing the emulsifying, spreading, dispersibility and/or wetting properties of liquids or solids.

☆ There are four major types of surfactants used in the herbicide formulation.

❖ Anionic surfactants are the alkaline metal salts of organic acids, *viz.* carboxylates, sulphonates, sulphates or phosphates and are most often used with acids or salts. The hydrophilic moieties in cationic surfactants are amino or quaternary nitrogen atoms, which form cations in aqueous phase.

❖ Cationic surfactants are useful as wetting agents, emulsifiers, dispersants, foam stabilizers, corrosion inhibitors, *etc.* Ethoxylated fatty amine, an adjuvant of this class, has been frequently used with the herbicide glyphosate.

❖ Non-ionic surfactants are neutral molecules and bear no electrical charge in solution and are generally compatible with most herbicides. The electronegative oxygen atom generates hydrophilicity.

❖ Amphoteric surfactant molecules bear both an acidic and a basic hydrophilic heads in the same molecule. Thus, they can produce both the ions, *i.e.* anions and cations, depending on the chemical nature of the environment. Non-ionic surfactants are neutral molecules and bear no electrical charge in solution and are generally compatible with most herbicides.

☆ The electronegative oxygen atom generates hydrophilicity. Uncharged or non-ionic surfactants are compatible with most herbicides. The major classes of non-ionic surfactants are alcohol ethoxylates, nonyl phenol ethoxylates, alkyl octylphenol ethoxylates, alkyl polysaccharides, urea clathrates, fatty acid ethoxylates, tallow amine ethoxyalates, phosphate esters, fatty acids, EO/PO block copolymers (organosilicones), and siloxanes (organosilicones).

☆ Binders and stickers are also important surfactants required for formulating herbicide technical. Binders are cementing materials required to hold the toxicant onto carrier. Some organic binders used in granular formulations are starch and dextrin, natural gum, shellac, resins, lignosulphonates and polyvinyl alcohol, plyethylene glycols, *etc.* Stickers are added in the formulation to improve its retention on the target surface.

☆ Polyethylene polysulphides, polyvinyl acetate, polybutenes, natural gums are some examples of stickers used in herbicide formulations. Spreader-stickers are the combinations of compounds that cause the droplet to spread over the surface and help in retaining the toxicants on the leaf.

☆ There are some special purpose adjuvants or utility adjuvants, which on addition in the formulation may widen the range of conditions under which a given herbicide is useful. They may alter the physical characteristics of the spray solution. This group includes compatibility agents, buffering agents, antifoam agents, and drift control agents.

☆ Compatibility agents allow simultaneous application of two or more ingredients. They are most often used when herbicides are applied in liquid fertilizer solutions. The addition of compatibility agents in the formulation may increase stability when mixing fertilizers and pesticides, emulsify spray solution and moderately adjusts pH for better tank mix component compatibility. Buffering agents usually contain a phosphate salt or more recently citric acid, which maintains a slightly acid pH when added to alkaline waters. These are added to higher pH solutions to prevent alkaline hydrolysis of some organophosphate (OP) and carbamate herbicides.

☆ Some buffering agents act as "water softening" agents that are used to reduce problems with hard water. In particular, calcium and magnesium salts may interfere with the performance of certain herbicides. Ammonium sulfate is sometimes added to reduce hard water problems.

Selection of Spray Adjuvant

☆ Adjuvant selection should be based on several factors including what the pesticide calls for, what the adjuvant claims to be, cost of the adjuvant, and what is available in your area. The primary source in deciding whether an adjuvant is necessary and the type of adjuvant used should come from the pesticide label.

✮ The following are some general guidelines to consider when given a choice of adjuvants.

  ❖ If both oil concentrate (crop or vegetable oil) and non-ionic surfactants are listed, then nonionic surfactant should be used under normal weather conditions when weeds are small and well within label guidelines. It is advised to use oil concentrate if weeds are stressed due to dry weather or with more mature weeds.

  ❖ If labeled, oil concentrate may be included in the formulation for control of grasses.

  ❖ Nitrogen fertilizer should only be used if it is recommended on the herbicide label.

  ❖ If the potential for crop injury is great, then nonionic surfactant should be included instead of oil concentrate.

  ❖ To improve crop safety, oil concentrates should not be used with plant growth regulator-type herbicides (*i.e.*, dicamba, 2,4-D, *etc.*).

✮ In ready-to-apply formulations, adjuvants are essential components along with toxicant. Nowadays, tank-mix adjuvant market is rapidly growing (Underwood, 2000). In most countries there is no patent protection or government regulation on the use of adjuvants in pesticide formulation. Manufacturers usually do not disclose the compositions of adjuvants in formulation to avoid others who can copy the composition of the non-proprietary items. Introduction of proprietary adjuvants is needed to have well defined composition for any herbicide formulation.

✮ In ready-to-apply formulations, adjuvants are essential components along with toxicant. Nowadays, tank-mix adjuvant market is rapidly growing (Underwood, 2000). In most countries there is no patent protection or government regulation on the use of adjuvants in pesticide formulation. Manufacturers usually do not disclose the compositions of adjuvants in formulation to avoid others who can copy the composition of the non-proprietary items. Introduction of proprietary adjuvants is needed to have well defined composition for any herbicide formulation.

Safener

✮ Herbicide safeners are chemical agents that selectively protect crop plants from herbicide damage without affecting the activity in target weed species. The concept to enhance crop tolerance to less selective or nonselective herbicide by using chemical agents was established by Otto Hoffman, when he accidentally observed the safening effect of 2,4-D in 2,4,6-T-treated tomato plants.

✮ The concept of using chemical safeners in the herbicide formulation practically started with the introduction of 1,8-naphthalic anhydride

(NA) to improve the tolerance of maize to thiocarbamate herbicides. Presently, many herbicides of different chemical classes are commercially used along with herbicide formulation. Safeners act in crop plants by reducing the uptake and transport of herbicide to reach the target site, or by interacting directly by inhibiting receptor proteins.

Commercially available Herbicide Safeners

Safener		Herbicide(s) Counteracted	Crop(s) Protected	Application Method
Chemical Class	Name			
Anhydride	1,8-naphthalic anhydride (NA)	Thiocarbamates (EPTC, Butylate, Venolate)	Maize	Seed treatment
Dichlcroace-tamide	Dicialormid	Thiocarbamates, Chloroacetanilide	Maize	Pre-plant incorporated with herbicide
	Furilazole	Acetochlor, Halosulfuron-methyl	Maize	Spray (Pre-em) as a mixture with herbicide
	AD-67	Acetochlor	Maize	Spray (Pre-em) as a mixture with herbicide
	Benoxacor	Metolachlor	Maize	Spray (Pre-em) as a mixture with herbicide
Oxime ether	Cyometrinil	Odoroacetanilide (Metolachlor)	Sorghum	Seed treatment
	Oxabetrinil	Chloroacetanilide (Metolachlor)	Sorghum	Seed treatment
	Fluxofenim	Ciloroacetanilide (Metolachlor)	Sorghum	Seed treatment
Thiazole carboxylic acid	Flurazole	Alachlor	Sorghum	Seed treatment
Dichloromethyl ketal	MG-191	Thiocarbamates, Chloroacetanilide	Maize	Spray (Pre -em) as a mixture with herbicide
Thenylpyri-midine	Fenclorim	Pretilachlor	Rice	Spray (Pre-cm) as a mixture with herbicide
Urea	Dymron	Pyributicarb, Pretilachlor, Prazosulfumn-ethyl	Rice	Spray (Pm-ern, Past - em) as a mixture with herbicide
Pmendow-1-carbothioate	Denemperate	Stafonylureas	Rice	Spray (Post-em) as mixture with herbicide

Safener		Herbicide(s) Counteracted	Crop(s) Protected	Application Method
Chemical Class	Name			
8-Quinolinoxy-carboxylic esters	Cloquintocet mexyl	Clcdinafop-Propergy	Cereals	Spray (Post-cm) as mixture with herbicide
1.2,4-Triazole-carboxylate	Fenchlorazole-ethyl	Fenoxaprop-ethyl	Cereals	Spray (Post -cm) as mixture with herbicide
Dihydro-pyrazole- dicar-boxylate	Mefenpyr-diethyl	ACCase inhibitors (Sulfonyl ureas)	Wheat Rye, Triticale, Barley	Spray (Post em) as mixture with herbicide
Dihydrois-oxazole- Cvarb-oxylate	Isoxactifert-ethyl	ACCase inhibitors (Sulfonyl ureas)	Maize, Rice	Spray (Past-em) as mixture with herbicide
Arylsulfonyl-benzarnide	Cyprosulfamide	Isoxaflutole	Maize	Spray (Pre-em, Post - em) as a mixture with herbicide.

Adopted from Stephenson and Yaaocoby (1991); Davies (2001) and Rosinger (2014).

Phytotoxicity of Herbicides

Herbicide Type	Symptoms
Pre-emergence herbicides	1. Reduce germination
	2. Suppresses crop growth
	3. Produces deformity in crop plants Post-emergence herbicides
Pre-emergence herbicides	1. Leaf injury
	2. Wilting
	3. Vein clearing
	4. Necrosis
	5. Epinasty
	6. Hyponasty
	7. Yellowing or chlorosis
	8. Sunting or scorching

Herbicides and their Respective Mode of Action

Herbicides	Mode of Action
1. IPC	Causes of epindal to boundry layer
2. 2,4-D ethyl ester	Highly volatile - Abnormal cell division
3. 2,4-D sodium salt	Highly soluble - Abnormal cell division
4. Glyphosate	Non selective, translocated and zero persistence and general metabolic inhibitors

Herbicides	Mode of Action
5. Diquate	Disturb of cell permeability
6. Triazines group	Photosynthesis inhibitor
7. Atrazine	Selective (Conjugation)
8. Pendimethaline	Microtubule assembly inhibition
9. Paraquate	Contact herbicide; inhibition of DNA synthesis
10. Dinitroaniline herbicides	Inhibition of respiration
11. Oxadiazone	Inhibition of CO_2
12. Trifluralin	Inhibition of RNA synthesis
13. Btachlor and Alachlor	Inhibition of protein synthesis and GA production during germination
14. Dalapan	Inhibition of vitamin synthesis
15. Thio-carbamate	Inhibition of lipid synthesis
16. Carbamate groups	Inhibition of cell division
17. Auxin type herbicides	Abnormal tissue development
18. Propanil	Degradation or metabolism
19. 2,4-DB	Reverse metabolism
20. Paraquate, Diquate and Glyphosate	Knock down effect

Mode of Action

Contact

Weed/Plant
Death

Absorption

Herbicide
Mode of
Action

Toxicity

Movement

Herbicide Mode of Action – Stages.

Examples of Herbicide Concentrations Causing Toxic Effects

Herbicide	Taxa	Biological Effect
Glyphosate	Water flea *Daphnia magna*	Acute 48h EC_{50} is 218 mg/L (ECOTOX)
	Amphipod *Gammarus pseudolimnaeus*	Acute 48h EC_{50} is 42-62 mg/L (ECOTOX)
	Buzzer midge *Chironomus plumosus*	Acute 48h EC_{50} is 55 mg/L technical glyphosate and 13mg/L Roundup® surfactant (Folmar *et al.*, 1979)
	Mayfly *Ephemerella walkeri*	Avoided Roundup® at 10 mg/L but not 1.0 mg/L (Folmar *et al.*, 1979)
	Channel catfish *Ictalurus punctatus*	Acute 96h LC_{50} is 130mg/L technical glyphosate and 13mg/L Roundup® surfactant (Folmar *et al.*, 1979)
	Fathead minnow *Pimephales promelas*	Acute 96h LC_{50} is 97mg/L technical glyphosate and 1.0 mg/L Roundup® surfactant (Folmar *et al.*, 1979)
	Rainbow trout *Oncorhynchus mykiss*	More sensitive response to Roundup® at elevated temperatures and at pH as it rises from 6.5 to 7.5, with no increased sensitivity at pH beyond 7.5 (Folmar *et al.*, 1979)
	Bluegill sunfish *Lepomis macrochiru*	
	American ribbed fluke snail *Pseudosuccinia columella*	Continuous exposure across generations produced reproductive effects on the third generation including rapid embryonic development, embryonic abnormalities and increased egg laying (Tate *et al.*, 1997)
Atrazine	Midge *Labrundinia pilosella*	Reduced emergence at 20 ug/L (Dewey 1986)
	Cream and brown microcaddisfly *Oxyethira pallida*	Shift in emergence period at 20 ug/L (Dewey 1986)
	Non-predatory insects	Reduced abundance at 20 ug/L (Dewey 1986)
	Stonewort algae *Chara* sp.	Resistant to atrazine up to 100 ug/L (Dewey 1986)
	Tiger salamander *Ambystoma tigrinum* sp.	Increased larval stage duration, reduced weight and body size (Larson *et al.*, 1998)
	Hydra sp.	48 hr LC_{50} of 3,000 ug/L (lowest acute value) (U.S. EPA 2003)
	Goldfish *Carassius auratus*	96 hr LC_{50} of 60,000 ug/L (highest acute value) (U.S. EPA 2003)
	Water flea *Ceriodaphnia dubia*	Life cycle chronic value of 3,536 ug/L (highest chronic value) (U.S. EPA 2003)
	Brook trout *Salvelinus fontinalis*	Life cycle chronic value of 88.32 ug/L (lowest chronic value) (U.S.EPA 2003)

Herbicide Resistance in Weeds

☆ Herbicide resistance is the genetic capacity of a weed population to survive a herbicide treatment that, under normal use conditions, would effectively control that weed population.

☆ Herbicide resistance is an example of evolution happening at an accelerated pace and an illustration of the "survival of the fittest" principle. A herbicide may kill all the weeds in a population of a particular weed species except for a few individuals with the genetic capacity to survive the herbicide.

Herbicide resistance is the inherited ability of a plant to survive and reproduce following exposure to a dose of herbicide that would normally be lethal to the wild plant. Resistance happens with the repeated use of the same herbicide, or herbicides with similar modes of action on a weed population. Resistant plants were already found, very infrequently, in the weed population before a herbicide was ever used.

☆ Herbicide resistant weeds are normally very rare in a weed population. Applying the same herbicide in the same field year after year will select for resistant plants. The resistant weeds set seed and may eventually dominate the population. This population is then not effectively controlled by the selecting herbicide.

Management

Some management techniques are key in preventing herbicide resistance.

☆ Use mechanical weed control methods, such as cultivation, to control weeds.

☆ Rotate herbicides—do not make more than two consecutive applications of herbicides with the same mode of action in the same field.

☆ Use tank mixtures of herbicides with differing effective modes of action.

☆ Rotate crops.

☆ Scout your fields and destroy weed escapes.

☆ Use herbicides with short soil residual times—herbicides with long soil residual times generally favor herbicide resistance.

☆ Clean your equipment before moving to a different field to prevent the spread of resistant biotypes and save work in fields with suspected herbicide resistance for last.

Chapter 12

Cropping Systems

Introduction

☆ Concept of cropping system is as old as agriculture in India. Multiplicity of cropping systems has been one of the main features of Indian agriculture and it is mainly attributed to prevailing socio-economic situations of farming community.

☆ The term cropping system essentially represents a philosophy of maximum crop production per unit area of land within a calendar year or relevant time unit with minimum natural resource degradation. Cropping systems remain dynamic in time and space, making it difficult to precisely determine their spread using conventional methods, over a large territory. However, it has been estimated that more than 250 double cropping systems are followed throughout the country.

☆ Based on rationale of spread of crops in each district in the country, 30 important cropping systems have been identified for irrigated conditions. These are; rice-wheat, rice-rice, rice-gram, rice-mustard, rice-groundnut, rice-sorghum, pearlmillet-gram, pearlmillet-mustard, pearlmillet-sorghum, cotton-wheat, cotton-gram, cotton-sorghum, cotton-safflower, cotton-groundnut, maize-wheat, maize-gram, sugarcane-wheat, soybean-wheat, sorghum-sorghum, groundnut-wheat, sorghum-groundnut, groundnut-rice, sorghum-wheat, sorghum-gram, pigeonpea-sorghum, groundnut-groundnut, sorghum-rice, groundnut-sorghum and soybean-gram (Das, 2010).

Cropping Systems

☆ Cropping systems, an important component of a farming system, represents a cropping pattern used on a farm and their interaction with farm resources, other farm enterprises and available technology, which determine their makeup.

Cropping systems is defined as the order in which the crops are grown or cultivated on a piece of land over fixed period.

☆ In the cropping systems, sometimes a number of crops are grown together or they are grown separately at short intervals in the same field.

Cropping Pattern

☆ It is the pattern of crops for a given piece of land or cropping pattern means the proportion of area under various crops

Cropping pattern is the yearly sequence and spatial arrangement of crops on a same piece of land over a same period of time.

at a point of time in a unit area or it indicated the yearly sequence and spatial arrangements of crops and follows in an area.

☆ Cropping pattern is a dynamic concept because it changes over space and time. It can be defined as the proportion of area under various crops at a point of time. In other words, it is a yearly sequence and spatial arrangement of sowing and fallow on a given area. In India, the cropping pattern determined by rainfall, climate, temperature, soil type and technology.

☆ Cropping pattern basically involves:

❖ Crop rotation practiced by a majority of the farmers in a given area or locality.

❖ Type and arrangement of crops in time and space.

❖ Yearly sequence and spatial arrangement of crops or of crops and fallow on a given area.

❖ The proportion of area under various crops at a point of time in the unit area.

Difference between Cropping Pattern and Cropping System

Cropping Pattern	Cropping System
Crop rotation practiced by a majority of farmers in a given area or locality	Cropping pattern and its management to derive benefits from a given resource base under specific environmental conditions.
Type and management of crops in time and space.	The cropping patterns used on a farm and their interaction with farm resources, other farm enterprises and available technology which determine their make up.

Cropping Pattern	Cropping System
Yearly sequence and spatial arrangement of crops or crops and fallow on a given area. The proportion of area under various crops at a point of time in a unit area	Pattern of crops taken up for a given piece of land, or order in which crops are cultivated on a piece of land over a fixed period, associated with soil, management practices such as tillage manuring and irrigation

Types of Cropping Systems

☆ The term cropping system refers to the crops, crop sequences and management techniques used on a particular agricultural field over a period of years.

☆ It includes all spatial and temporal aspects of managing an agricultural system.

☆ The Indian agriculture is decided by the soil types and climatic parameters which determine overall agro-ecological setting for nourishment and appropriateness of a crop or set of crops for cultivation.

☆ There are three distinct crop seasons in India, namely Kharif, Rabi and Zaid.

☆ The Kharif season started with Southwest Monsoon under which the cultivation of tropical crops such as rice, cotton, jute, jowar, bajra and tur are cultivated.

☆ The Rabi season starts with the onset of winter in October-November and ends in March-April.

☆ Zaid is a short duration summer cropping season beginning after harvesting of Rabi crops.

Types of Cropping System

System	Definition	Remarks
Monocropping	The continuous growing of the same species on a piece of land over a sequence of growing seasons	Risk of residual transfer of pests, diseases, and weeds from one season to the next
Ratooning	The residual stumps of a crop are allowed to regrow to produce a second crop	Cheap, low-input second crop with shorter growing period than sequential cropping. Yields of second crop are often poor and unreliable, often because of pest and disease transfer. Restricted to crops such as sorghum, millet, rice, and pigeonpea
Sole crop	Crop composed of individual plants of the same variety of one species	Interactions occur between plants that are all at the same stage of growth. Management of inputs, control measures, and irrigation can be synchronized to meet the temporal and spatial demands of the crop. Mechanization is possible

System	Definition	Remarks
Sequential cropping	Crops are grown one after the other with no overlapping phase	Usually involves an additional crop after the harvest of the main single rainy season crop. Potential residual effects in the reservoir of nutrients, weeds, or pests and diseases with implications for the succeeding crop or crops. Requires a relatively long potential growing period (over 180–200 days). The growth of the second crop is often risky because of the limited amount of residual moisture. Typically includes legumes or oilseeds after paddy rice or rainfed cereals
Multicropping	Two or more species growing on the same piece of land, where at least part of the growth cycles of different species overlap	Plants of different species are close enough to allow interactions between them, usually at different stages of their growth
Intercropping	A multicrop composed of two or more annual crops grown simultaneously on a piece of land within a single growing season	Intercrops may retain the same population as the sole crops that they are composed of in which each plant of one species is replaced by a single plant of one or more other species ("replacement intercrop") or additional plants of one or more species may be fitted into an existing population of sole crops ("additive intercrop")
Mixed intercrop	Multicrops of randomly arranged plants of different species	Such crops may be planted as replacement or additive intercrops where the grower requires an additional output from a second species without affecting the yield of the main crop
Row intercrop	Intercrops grown in structured arrangements of different species arranged in alternating blocks of narrow rows	Interspecies interactions occur between rows and intrarow competition occurs within rows. Individual species may be sown, fertilized, and harvested separately
Strip intercrop	Intercrops where blocked rows of each species are grown separately	Strips are wide enough for some plants within each block to behave as if in a sole crop
Relay cropping	The interplanting of one species with a second species before the first crop reaches maturity	The very short overlap means competition between species is minimal. The second crop may have better soil moisture conditions because of earlier sowing and protection from the first crop. Sowing of second crop before harvest of first crop reduces labor peaks at harvest. There may be damage to seedlings of second crop at harvest of first. Examples include paddy rice and legumes, cereals and legumes, cereals and other cereals

System	Definition	Remarks
Agroforestry	Land use systems in which woody perennials (trees, shrubs) are grown in association with herbaceous plants (annual or perennial crops, or pastures) and/or livestock	Capture and use of resources is distributed in space and time. Plants within the system make demands on resources at different times ("temporal complementarity") or use resources more efficiently at any one time ("spatial complementarity"). In agroforestry systems the woody component/s may provide more than one product (*e.g.*, timber, fuelwood, fodder, fruit) and/or "services" (shelter, shade, soil or water conservation)
Alley cropping	A specific form of agroforestry in which trees and shrubs are established in hedgerows within arable cropped land	The woody component is always planted in rows but the spacing between rows depends on the species mix and topography. Sometimes known as "hedgerow intercropping"

Type of Cropping Systems

Monocropping

- ✰ Monocropping is the agricultural practice of growing a single crop year after year on the same land. Rice, Maize, soybeans, and wheat are common crops often grown using monocropping techniques.

- ✰ Monocropping allows for farmers to have consistent crops throughout their entire farm. They can plant only the most profitable crop on their entire farm, which may increase overall farm profitability.

- ✰ It allows a farmer to specialize in a particular crop, which means that he or she can invest in machinery designed specifically for that crop to generate a large volume of the crop at harvest.

- ✰ It severely depletes the soil, as the plant will strip the soil of the nutrients it needs.

- ✰ This forces farmers to use fertilizers, which can disturb the natural balance of the soil and contribute to a host of environmental problems, from pollution to desertification.

- ✰ The practice can also contribute to the proliferation of crop pests and diseases, which can be a serious liability when a farmer's land is planted exclusively with one crop.

- ✰ Monocropping also generally reduces crop diversity.

- ✰ If a crop does become subject to a particular pest or disease, it becomes especially vulnerable. In a world where only a few strains of corn are grown, for example, if a pest develops to attack one, it could devastate global crops, and farmers might not have another strain to fall back upon.

☆ Additionally, the practice is very dangerous when natural disasters or shifting weather devastate a crop. Farmers may find themselves heavily in debt at the end of the season, and the lack of harvest could translate into famine or general hardship.

Monocropping

Intercropping

☆ Intercropping is a farming method that involves growing two or more crops together at the same time and on the same piece of land.

☆ There are at least four types of intercropping according to spatial arrangement.

❖ **Row Intercropping** is the growing of two or more crops at the same time with at least one crop planted in rows. In farms grown to perennial crops, annual crops like corn, rice and pineapple are commonly grown as intercrop between the rows of the main crop.

❖ **Strip Intercropping** is the growing of two or more crops together in strips wide enough to allow separate production of crops using mechanical implements, but close enough for the crops to interact.

❖ **Mixed Intercropping or Mixed Cropping** is the growing of two or more crops at the same time with no distinct row arrangement.

❖ **Relay Intercropping or Relay Cropping** is a system in which a second crop is planted into an existing crop when it has flowered (reproductive stage) but before harvesting. There is thus a minimum temporal overlap of two or more crops.

☆ It helps to suppress weeds since the crops take up much space that would have allowed the weeds to grow. Some weeds also find it difficult to grow alongside some crops.

☆ Growing two crops alongside each other can be of great benefit, especially if their interactions increase the fitness of one or both plants. For instance, plants that are likely to tip over in wind may gain structural support from their companions. Some plants may also provide shade to the light-sensitive plants.

☆ Pests can be controlled through intercropping by trap cropping, repellant intercropping, or push-pull cropping.

☆ A possible problem is that the intercrop may compete with the main crop for light, water and nutrients. This may reduce the yields of both crops.

Sequential Cropping

☆ This involves growing two crops in the same field, one after the other in the same year.

☆ In some places, the rainy season is long enough to grow two crops: either two main crops, or one main crop followed by a cover crop.

☆ Growing two crops may also be possible if there are two rainy seasons, or if there is enough moisture left in the soil to grow a second crop. If the crops are different, this is a crop rotation.

Crop Rotation

☆ This means changing the type of crops grown in the field each season or each year (or changing from crops to fallow).

☆ Crop rotation is a key principle of conservation agriculture because it improves the soil structure and fertility, and because it helps control weeds, pests and diseases.

☆ Crop rotation is one of the oldest and most effective cultural control strategies. It means the planned order of specific crops planted on the same field.

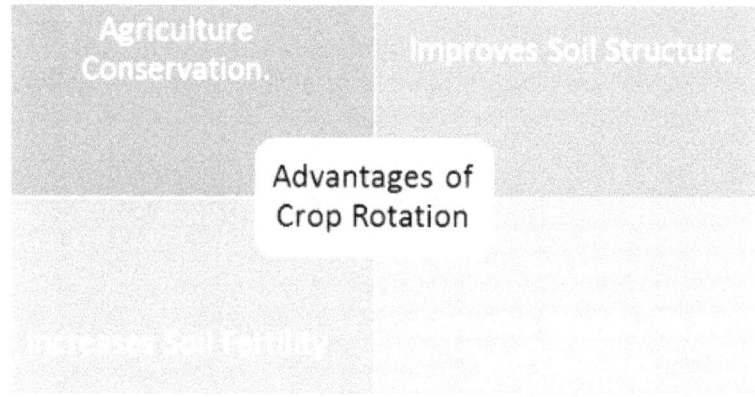

★ It also means that the succeeding crop belongs to a different family than the previous one. The planned rotation may vary from 2 or 3 year or longer period.

Alley Cropping System

★ Alley cropping system is the cultivation of food, forage, or specialty crops between rows of trees.

★ This system is a larger version of intercropping or companion planting conducted over a longer time scale.

★ In the Alley cropping system, rows of trees are planted at wide spacing with a companion crop grown in the Alleyways between the rows.

★ Alley cropping improves farm income, crop production, and protects crops.

★ It allows the farmer to effectively use available resources and yield more benefits.

★ One main disadvantage of the Alley cropping system is that additional labour is required to prune the trees.

Cropping Pattern in India

★ A broad picture of the major cropping patterns in India can be prevented by taking the major crops into consideration.

★ The cropping pattern of a particular area can be found in two steps:

❖ The crop occupying the highest percentage of the sown area of the region is taken as the base crop.

❖ All other possible alternative crops which is sown in the region is considered in the pattern.

❖ Example: In the given figure, if we see the area of Maharashtra we have cotton as the base crop, but jowar is also grown in Maharashtra on large scale, hence jowar is an alternative crop in Maharashtra making cotton-jowar as the cropping pattern. Similarly in the eastern states of Bengal, Assam, Orissa, rice is the base crop but jute and tea are also grown at a large scale thus making rice-jute–tea as the cropping pattern in eastern India.

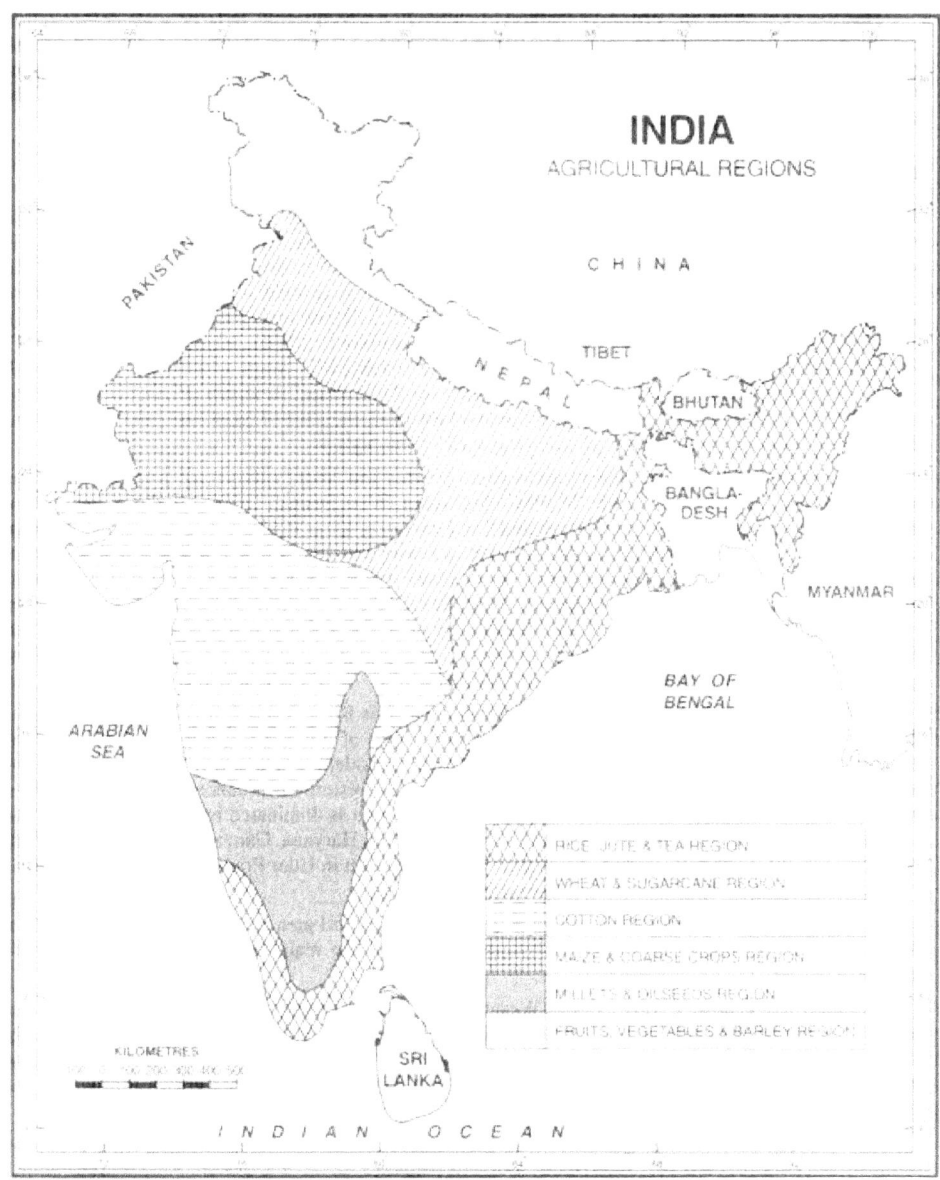

Agricultural Regions of India.

Factors Determining Cropping Pattern In India

1. Physical Factors

☆ Cropping pattern of any particular region of the country is depending on its soil content, weather, climate, rainfall *etc.*

☆ As for example, in a wet area having chances of heavy rainfall and water-logging, people will like to cultivate rice whereas in a dry area, farmer can manage to cultivate coarse cereals like bajra, jowar *etc.*

2. Technical Factors

☆ The cropping pattern also depend upon the technical factors such as nature and capacity of irrigation facilities available in a region, availability of improved seeds, chemical fertiliser *etc.* With the development of irrigation facilities, the entire method of cultivation being followed from the traditional period is bound to change.

☆ With this, new and better crop rotation system can be followed and new and superior crops also can be grown.

☆ In India, due to the extension of irrigation facilities, the cultivation of sugarcane, tobacco, oilseeds *etc.* have increased substantially.

☆ Moreover, with the availability of irrigation water, even double or triple cropping is also successfully done. Again, in the absence of irrigation facilities in some other parts of the country, the concept of "dry land farming" is also gaining its importance in recent year.

3. Economic Factors

☆ Economic factors are playing the major role in determining the cropping pattern in a country like India.

☆ The following are some of the economic factors influencing the cropping pattern of our country:

❖ *Price and income aspect:* Movement of price of agricultural products is having some correlation with the changes in cropping pattern. A remunerative and steady price of a particular crop will provide a better incentive to the producer to produce that crop and un-remunerative price will induce the farmer to change the cropping pattern In India, fixed procurement price of wheat and rice and other controls imposed by the Government induced the farmers to shift to cash crops like sugarcane. Again, the unremunerative prices of jute prevailing in Assam and other adjoining states also led to shift in the production of food crops. Moreover, income maximisation aspect is also playing an important role in influencing the cropping pattern in the country. Relative profitability per acre is also having considerable influence on the cropping pattern of the country.

❖ **Farm Size:** A good relationship also exists between farm size and cropping pattern. In a small farm, farmers are very much interested to produce food grains for household consumption. After meeting their own food requirements small farmers may go for cash crops in order to maximise their money income. On the other hand, in a big farm farmers like to follow that cropping pattern which maximise their income.

❖ **Tenure:** Land tenure system prevailing in the country also influences the cropping pattern. In a system of crop sharing, it is the landlord who finalizes the cropping pattern guided by profit maximising principle.

❖ **Availability affirm inputs:** Cropping pattern is also depending upon the farm inputs available, seeds, fertiliser, controlled and assured water supply through irrigation *etc.* and among these irrigation is the most important.

❖ **Government Action:** Cropping pattern may also be influenced by government action undertaken in the form of administrative and legislative measure. Supply of inputs by the government, intensive scheme for various crops, various government campaign like grow more food or any legislative provision by the government, transportation and marketing provision also help to finalize the cropping pattern in the country.

Dryland Farming

✫ Dryland farming is cultivation of crops in regions with annual rainfall more than 750 mm. In spite of prolonged dry spells crop failure is relatively less frequent. These are semiarid tracts with a growing period between 75 and 120 days. Moisture conservation practices are necessary for crop production. However, adequate drainage is required especially for vertisols or black soils.

✫ Dryland farming is agriculture dependent upon the vagaries of weather, especially precipitation. In its broadest aspects, dryland farming is concerned with all phases of land use under semiarid conditions. Not only how to farm but how much to farm and whether to farm must be taken into consideration. Above all else, dryland farming must emphasize the capture and efficient use of precipitation.

✫ Rainfed farming and dryland farming are often used interchangeably, but this is a serious error. They both exclude irrigation, but beyond that, they can differ significantly. Dryland farming is a special case of rainfed agriculture practiced in arid and semiarid regions in which annual precipitation is about 20–35 per cent of potential evapotranspiration. Conditions of moderate-to-severe moisture stress occur during a substantial part of the year, greatly limiting yield potential, and in which farming emphasizes water conservation in all practices throughout the year.

☆ Rainfed systems, although they include dryland systems, can also include systems which emphasize disposal of excess water, maximum crop yields, and high inputs of fertilizer. There are three components of a successful dryland farming system: (1) retaining the precipitation on the land, (2) reducing evaporation from the soil surface to increase the portion of evapotranspiration used for transpiration, and (3) utilizing crops that have drought tolerance and that fit the precipitation patterns. Although these components have been known for centuries, new technologies continue to be developed that increase crop production in water-short areas.

Benefits of Dryland

The primary benefit is obvious – the ability to grow crops in arid regions without supplemental irrigation. In this day and age of climate change, the water supply is becoming increasingly precarious. This means that farmers (and many gardeners) are looking for new, or rather old, methods of producing crops. Dryland farming might just be the solution.

Crops Grown in Dryland Farming

At one point, a variety of crops were produced using dryland farming methods. As mentioned, there is a renewed interest in dry farming crops. Research is being done on (and some farmers are already utilizing) dry farming of dry beans, melons, potatoes, squash, and tomatoes.

Chapter 13

Factors Affecting Crop Production

Crop

Crop refers to plants sown and harvested by man for economic use. A crop is a non-animal species or variety that is grown to be harvested as food, fodder, timber, fuel or for any other economic purpose.

Crop Production

It ideals with the production of various crops, which includes food crops, fodder crops, fiber crops, sugar, oilseeds *etc.* It includes agronomy, soil science, entomology, pathology, microbiology *etc.* Crop production is a complex process. Its success depends on crop and environmental factors both. Apart from these, socio-economic and political factors also determine the success of crop production. Crop production as an art, science and business.

Plant Growth

☆ Growth is the permanent, irreversible increase in the size of an organism. This feature is observed in all organisms, accompanied by several metabolic processes.

☆ In plants, the seeds germinate and develop into a new seedling, which finally develops into an adult plant. Plants display indefinite growth.

Phases of Plant Growth

These are the phases of plant growth:

Formative Phase

Plants grow by cell division. The pre-existing cells divide to give rise to new cells. The process of cell division in plants is known as mitosis. It is carried out in two steps:

⭐ Division of Nucleus or Karyokinesis

⭐ Division of Cytoplasm or Cytokinesis

In higher plants, the division of cells begins in the meristematic region.

Cell Enlargement and Differentiation

The size of the cells, tissues and organs increases at this stage by the formation of protoplasm, absorption of water, developing vacuoles, and addition of cell walls to make it thicker and permanent.

Cell Maturation

The enlarged cells acquire a definite shape and form at this stage. This helps in differentiating different cells and tissues.

Factors Affecting Crop Production

Several factors affected the crop production. These factors are to be taken into consideration by a crop producer (farmer) during crop planning and crop production. Some of these factors are under control of a producer, whereas, others can be modified for obtaining good results. On the contrary, some factors are beyond the control of crop producer.

⭐ Factor within total control?

⭐ Factors that can be manipulated

⭐ Factors outside the producer's control

⭐ Elements affecting crop production

There are a variety of factors associated with crop yield and the risks involved with farming. Factors affecting crop production – climatic – edaphic - biotic- physiographic and socio economic factors

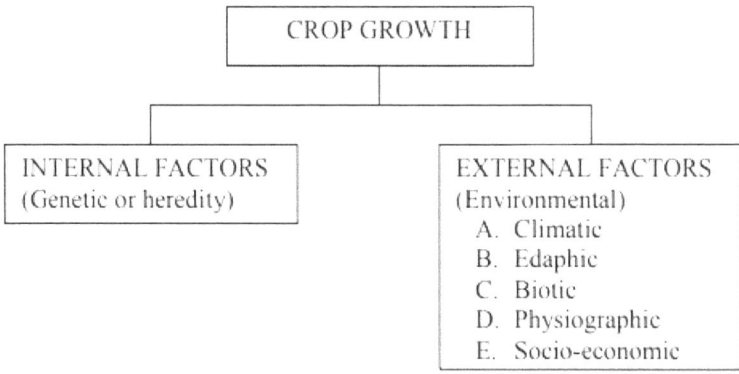

1. Internal Factors

Genetic Factors

The genetic factor determines the character of a plant, but the extent to which this is expressed is influenced by the environment. The genetic factor is also called internal factor because the basis of plant expression (the gene) is located within the cell. The increase in crop yields and other desirable characters are related to Genetic makeup of plants.

The genotype of a plant affects its growth. For example, selected varieties of rice grow rapidly, maturing within 110 days, whereas others, in the same environmental conditions, grow more slowly and mature within 155 days. A producer has control over the genetic factor by his choice of variety.

High yielding ability, Early maturity, Resistance to lodging, Drought flood and salinity tolerance, Tolerance to insect pests and diseases, Chemical composition of grains (oil content, protein content), Quality of grains (fineness, coarseness) and Quality of straw (sweetness, juiciness). These characters are less influenced by environmental factors since they are governed by genetic make-up of crop.

2. External Factors

The environmental factor is considered external, and refers to all factors, climatic, edaphic, biotic physiographic and socio-economic other than the genetic factor.

A. Climatic Factors

☆ Nearly 50 per cent of yield is attributed to the influence of climatic factors. In nature, there is an interaction between these factors and they all affect each other. In a controlled environment, such as a nursery or open field seed bed, temperature is the most influential factor in this interaction. A plant has the natural ability to regulate its level of activity according to environmental conditions, such as at specific levels of temperature and humidity. At extreme temperatures and humidity, such as when it is extremely dry or humid or extremely hot or cold, growth will stop, which may lead to the plant dying if the conditions persist.

☆ Environmental conditions therefore play an important role in the ability of a plant to grow and in general plant health. Effectively controlling these factors enables one to propagate and grow healthy plants. The atmospheric weather variables which influence the crop production are precipitation, temperature, atmospheric humidity, solar radiation, wind velocity and atmospheric gases.

1. Precipitation

☆ **Precipitation,** all liquid and solid water particles that fall from clouds and reach the ground. These particles include drizzle, rain, snow, snow pellets, ice crystals, and hail. Precipitation is water released from clouds in the form of rain, freezing rain, sleet, snow, or hail. It is the primary connection in the water cycle that provides for the delivery of atmospheric water to the Earth. Most precipitation falls as rain.

☆ Precipitation includes all water which falls from atmosphere such as rainfall, snow, hail, fog and dew. Rainfall one of the most important factor influences the vegetation of a place. Total precipitation in amount and distribution greatly affects the choice of a cultivated species in a place. In heavy and evenly distributed rainfall areas, crops like rice in plains and tea, coffee and rubber in Western Ghats are grown. Low and uneven distribution of rainfall is common in dryland farming where drought resistance crops like pearl millet, sorghum and minor millets are grown. In desert areas grasses and shrubs are common where hot desert climate exists. Though the rainfall has major influence on yield of crops, yields are not always directly proportional to the amount of Precipitation as excess above optimum reduces the yields. Distribution of rainfall is more important than total rainfall to have longer growing period especially in drylands.

2.   Temperature

☆ If heat and light, which cause an increase in temperature, is not controlled properly, plants may suffer from heat injury. The ideal temperature for propagation is 29°C, and it must be monitored closely. In propagation chambers, the temperature can often be maintained at this ideal level by heating and cooling systems. The heat is also used for increasing the humidity in the chambers, by drenching the trays and dampening the floor.

☆ Temperature is a measure of intensity of heat energy. The range of temperature for maximum growth of most of the agricultural plants is between 15 and 40°C. The temperature of a place is largely determined by its distance from the equator (latitude) and altitude. It influences distribution of crop plants and vegetation. Germination, growth and development of crops are highly influenced by temperature. Affects leaf production, expansion and flowering.

☆ Physical and chemical processes within the plants are governed by air temperature. Diffusion rates of gases and liquids changes with temperature. Solubility of different substances in plant is dependent on temperature. The minimum, maximum (above which crop growth ceases) and optimum temperature of individual's plant is called as cardinal temperature.

Crops	Minimum Temperature °C	Optimum Temperature °C	Maximum Temperature °C
Rice	10	32	36-38
Wheat	4.5	20	30-32
Maize	8-10	20	40-43
Sorghum	12-13	25	40
Tobacco	12-14	29	35

3. Atmospheric Humidity (Relative Humidity - RH)

☆ Humidity, also referred to as relative humidity, is the amount of water vapour in the air at a given temperature, and is expressed as a percentage. This means that at 20 per cent relative humidity, 20 per cent of any given volume of air will consist of suspended water molecules. Humidity levels are especially important in allowing the plant to carry on with its metabolic processes at desired rates. The ideal relative humidity for propagation ranges between 80 per cent and 95 per cent for seeds and cuttings, and in the region of 60 per cent outdoors for budding, grafting and seedbed methods. Seed germination is faster at higher humidity levels, as is the case of cuttings. In warm and dry areas, the level of humidity often falls below 55 per cent on hot summer days, making budding and grafting more delicate and requiring close monitoring.

☆ Water is present in the atmosphere in the form of invisible water vapour, normally known as humidity. Relative humidity is ratio between the amounts of moisture present in the air to the saturation capacity of the air at a particular temperature. If relative humidity is 100 per cent it means that the entire space is filled with water and there is no soil evaporation and plant transpiration. Relative humidity influences the water requirement of crops. Relative humidity of 40-60 per cent is suitable for most of the crop plants. Very few crops can perform well when relative humidity is 80 per cent and above. When relative humidity is high there is chance for the outbreak of pest and disease.

4. Solar Radiation (without which life will not exist)

☆ All green plants require light for growth to take place. Some plants (most species) prefer growing in direct sunlight, while others prefer growing in the shade where they are subjected to indirect sunlight. Light is essential for photosynthesis, while light quality, which is determined by the wavelength of the light, also influences germination and flowering. Plants grown under protection such as greenhouses and shade houses, require adequate light for the process of photosynthesis. If the plant does not receive enough light, which may be due to shading or over-crowding, it displays symptoms of retarded growth.

☆ In seed germination, red light, with a wavelength of 660 nanometre (nm), is used in chambers to stimulate germination of certain kinds of seeds. Incandescent globes are commonly used as an artificial source for red light for this purpose, while fluorescent tubes provide the blue light required for photosynthesis after germination. These lights are used extensively and kept on for as long as possible. It is not uncommon to have lights on 24 hours a day week round. The depth of sowing light sensitive seed also determines the time seeds take to germinate, because light cannot penetrate deeply into the soil. Therefore light sensitive seeds should be planted shallower than non-sensitive seeds. With no or inadequate light, weak seedlings of poor quality are produced. These seedlings display an excessive elongation, referred to as etiolation.

☆ From germination to harvest and even post harvest crops are affected by solar radiation. Biomass production by photosynthetic processes requires light. All physical process taking place in the soil, plant and environment are dependent on light. Solar radiation controls distribution of temperature and there by distribution of crops in a region. Visible radiation is very important in photosynthetic mechanism of plants. Photo-synthetically Active Radiation (PAR - 0.4 – 0.7μ) is essential for production of carbohydrates and ultimately biomass.

0.4 to 0.5 μ - Blue – violet – Active

0.5 to 0.6 μ - Orange – red – Active

0.5 to 0.6 μ - Green –yellow – low active

☆ Photoperiodism is a response of plant to day length. Short day – Day length is 12 hours (Barley, oat, carrot and cabbage), day neutral – There is no or less influence on day length. Phototropism –– Response of plants to light direction. *e.g.* Sunflower. Photosensitive – Season bound varieties depends on quantity of light received.

5. Wind Velocity

☆ Wind velocity, is a fundamental atmospheric quantity caused by air moving from high to low pressure, usually due to changes in temperature. Note that wind direction is usually almost parallel to isobars (and not perpendicular, as one might expect), due to rotation. Wind velocity affects weather forecasting, aviation and maritime operations, construction projects, growth and metabolism rate of many plant species, and has countless other implications.

☆ The basic function of wind is to carry moisture (precipitation) and heat. The moving wind not only supplies moisture and heat, also supplies fresh CO_2 for the photosynthesis. Wind movement for 4 – 6 km/hour is suitable for more crops. When wind speed is enormous then there is mechanical damage of the crops (*i.e.*) it removes leaves and twigs and damages crops like banana, sugarcane. Wind dispersal of pollen and seeds is natural and necessary for certain crops. Its cause's soil erosion helps in cleaning produce to farmers, increases evaporation and spread of pest and diseases.

6. **Atmospheric Gases on Plant Growth**

   ☆ Atmospheric gases are gases located in the Earth's atmosphere. These gases primarily include oxygen and nitrogen (making up 99 per cent of the air); though greenhouse gases (carbon dioxide, methane, nitrous oxide, water vapor and ozone) make up 1 per cent of the air. CO_2 – 0.03 per cent, O2 - 20.95 per cent, N_2 - 78.09 per cent, Argon - 0.93 per cent, Others - 0.02 per cent. CO_2 is important for Photosynthesis, CO2 taken by the plants by diffusion process from leaves through stomata.

   ☆ CO_2 is returned to atmosphere during decomposition of organic materials, all farm wastes and by respiration. O_2 is important for respiration of both plants and animals while it is released by plants during Photosynthesis. Nitrogen is one of the important major plant nutrient, Atmospheric N is fixed in the soil by lightning, rainfall and N fixing microbes in pulses crops and available to plants. Certain gases like SO_2, CO, CH_4, HF released to atmosphere are toxic to plants.

B. Edaphic Factors (Soil)

   ☆ The edaphic factor includes the physical, chemical, and biological properties of soil that result from biologic and geologic phenomena or anthropogenic activities. Chemical and physical features of soil greatly influence the ecology and evolution of plants and their associated biota.

   ☆ Plants grown in land completely depend on soil on which they grow. The soil factors that affect crop growth are soil moisture, soil air, soil temperature, soil mineral matter, soil organic matter, soil organisms and soil reactions.

1. Soil Moisture

Moisture is essential for germination and healthy plant growth. Too much water suffocates the plant roots, and can cause diseases such as root rot, damping off, and collar rot. The other extreme is insufficient water supply, or drought, and is detrimental to all plants, but even more so to cuttings and young seedlings. A uniform and constant water supply is required for seed germination to produce healthy and vigorous seedlings, and for seedlings to grow into healthy plants. In all propagation methods, the properties of the growth medium determine the quality and quantity of water that will be available for uptake by the plant. A good medium is one that has a low salinity level, sufficient water holding capacity (50-60 per cent), make water available to the plant easily, and the ability to allow lateral water movement. In the case of germination, the seed, and the later seedling stage, has to be kept in media wetted to field capacity, being the maximum amount of water that a particular soil can hold.

✰ Water is a principal constituent of growing plant which it extracts from soil

✰ Water is essential for photosynthesis

✰ The moisture range between field capacity and permanent wilting point is available to plants.

✰ Available moisture will be more in clay soil than sandy soil

✰ Soil water helps in chemical and biological activities of soil including mineralization

✰ It influences the soil environment *e.g.* it moderates the soil temperature from extremes

✰ Nutrient availability and mobility increases with increase in soil moisture content

2. Soil Air

In nutrient management, soil aeration influences the availability of many nutrients. Particularly, soil air is needed by many of the microorganisms that release plant nutrients to the soil. An appropriate balance between soil air and soil water must be maintained since soil air is displaced by soil water.

✰ Aeration of soil is absolutely essential for the absorption of water by roots

✰ Germination is inhibited in the absence of oxygen

✰ O_2 is required for respiration of roots and micro organisms.

✰ Soil air is essential for nutrient availability of the soil by breaking down insoluble mineral to soluble salts

✰ For proper decomposition of organic matter

✰ Potato, tobacco, cotton linseed, tea and legumes need higher O_2 in soil air

✰ Rice requires low level of O_2 and can tolerate water logged (absence of O_2) condition.

3. Soil Temperature

Temperature of the soil has significant effect on the growth and development of plants mainly through its action on the absorption of water and minerals. Low temperature decreases the rate of respiration in the embryonic cells of the root and thereby checking its elongation, resulting in a slower rate of penetration into new areas of soil where water is available. A plant growing in soil saturated with water may wilt if the temperature of the soil falls below a certain degree, because at a very low temperature roots cannot absorb water from the soil. This effect is primarily due to increased viscosity of both water and protoplasm at low temperature.

Uptake of minerals is also greatly affected at low temperature because the reduced respiration results in less available energy during the absorption process and probably also because of great viscosity of the protoplasm. A certain degree of heat is necessary for seed germination, root growth and microbiological activity in the soil. All these activities almost cease at or near the freezing point of water. The temperature needed for seed germination and root growth vary with species to species. Microbiological activity is retarded by low soil temperature. As a result, the nitrification processes in the soil are slowed down and plant nutrition and growth are affected adversely. Plants growing on cold soils mostly show prostrate growing habit, whereas plants of the warm soils are usually slender and tall. Direct radiation from the sun, the heat generated by the decomposition of organic matter in the soil, and the heat from the earth's interior are the chief sources of soil heat. The temperature of the soil is affected by its colour, texture, slope and water content. Dark coloured soils absorb more heat than those of lighter hue. Sandy soils absorb heat during the day and lose it at night quicker than the finer grained silt and clay.

☆ It affects the physical and chemical processes going on in the soil.

☆ It influences the rate of absorption of water and solutes (nutrients)

☆ It affects the germination of seeds and growth rate of underground portions of the crops like tapioca, sweet potato.

☆ Soil temperature controls the microbial activity and processes involved in the nutrient availability

☆ Cold soils are not conducive for rapid growth of most of agricultural crops

4. Soil Mineral Matter

Mineral matter is the predominant component of mineral soils. It constitutes about 45 per cent of total soil composition. It is made up of a number of particles which vary in size, shape and chemical composition. These particles range from the microscopic colloidal clays to the coarse fraction of sand and gravel. In nature, mineral matter exists as a dispersed phase, implying that it can be separated into the individual particles that comprise it. Along with soil organic matter, it makes up the solid materials phase of the soil.

☆ The mineral content of soil is derived from the weathering of rocks and minerals as particles of different sizes.

☆ These are the sources of plant nutrients *e.g.* Ca, Mg, S, Mn, Fe, K *etc.*

5. Soil Organic Matter

On the basis of organic matter content, soils are characterized as mineral or organic. Mineral soils form most of the world's cultivated land and may contain from a trace to 30 percent organic matter. Organic soils are naturally rich in organic matter principally for climatic reasons. Although they contain more than 30 percent organic matter, it is precisely for this reason that they are not vital cropping soils.

✮ It supplies all the major, minor and micro nutrients to crops

✮ It improves the texture of the soil

✮ It increases the water holding capacity of the soil,

✮ It is a source of food for most microorganisms

✮ Organic acids released during decomposition of organic matter enables mineralisation process thus releasing unavailable plant nutrients

6. Soil Organisms

Soil organism, any organism inhabiting the soil during part or all of its life. Soil organisms, which range in size from microscopic cells that digest decaying organic material to small mammals that live primarily on other soil organisms, play an important role in maintaining fertility, structure, drainage, and aeration of soil. They also break down plant and animal tissues, releasing stored nutrients and converting them into forms usable by plants.

✮ Some soil organisms are pests. Among the soil organisms that are pests of crops are nematodes, slugs and snails, symphylids, beetle larvae, fly larvae, caterpillars, and root aphids. Some soil organisms cause rots, some release substances that inhibit plant growth, and others are hosts for organisms that cause animal diseases.

✮ The raw organic matter in the soil is decomposed by different micro organisms which in turn releases the plant nutrients. Atmospheric nitrogen is fixed by microbes in the soil and is available to crop plants through symbiotic (Rhizobium) or non-symbiotic (Azospirillum) association.

7. Soil Reaction (pH)

Soil pH or soil reaction is an indication of the acidity or alkalinity of soil and is measured in pH units. Soil pH is defined as the negative logarithm of the hydrogen ion concentration. The pH scale goes from 0 to 14 with pH 7 as the neutral point. As the amount of hydrogen ions in the soil increases the soil pH decreases thus becoming more acidic. From pH 7 to 0 the soil is increasingly more acidic and from pH 7 to 14 the soil is increasingly more alkaline or basic.

✮ The effect of soil pH is great on the solubility of minerals or nutrients. Fourteen of the seventeen essential plant nutrients are obtained from the soil. Before a nutrient can be used by plants it must be dissolved in the soil solution. Most minerals and nutrients are more soluble or available in acid soils than in neutral or slightly alkaline soils.

✮ Soil reaction is the pH (hydrogen ion concentration) of the soil. Soil pH affects crop growth and neutral soils with pH 7.0 are best for growth of most of the crops and Soils may be acidic (7.0). Soils with low pH are injurious to plants due high toxicity of Fe and Al and low pH also interferes with availability of other plant nutrients.

C. Biotic Factors

Biotic factors have their origin in the activities of living organisms, such as green and non-green plants, and all animals, including man. The activities of these living organisms have profound direct and indirect effects upon growth, structure, reproduction and distribution of plants on the earth. These effects result from the biotic relationships between the plants themselves comprising a plant community, between these plants and animals living in close proximity, and between the micro-fauna and flora of the soil. Biotic factors such as pests, insects and diseases reduce the crop production. A pest causes damage to our crops by feeding. Weeds also reduce crop productivity by competing with the main crop for nutrients and light. Beneficial and harmful effects caused by other biological organism (plants and animals) on the crop plants.

1. Plants

☆ Competitive and complementary nature among field crops when grown together. Competition between plants occurs when there is demand for nutrients, moisture and sunlight particularly when they are in short supply or when plants are closely spaced

☆ When different crops of cereals and legumes are grown together, mutual benefit results in higher yield. Competition between weed and crop plants as parasites *e.g.* Striga parasite weed on sugarcane crop.

2. Animals

☆ Soil fauna like protozoa, nematode, snails, and insects help in organic matter decomposition, while using organic matter for their living. Insects and nematodes cause damage to crop yield and considered as harmful organisms.

☆ Honey bees and wasps help in cross pollination and increases yield and considered as beneficial organisms. Burrowing earthworm facilitates aeration and drainage of the soil as ingestion of organic and mineral matter by earthworm results in constant mixing of these materials in the soils. Large animals cause damage to crop plants by grazing.

D. Physiographic Factors

☆ Physiographic factors of the habitat include form, behavior and structure of the earth's surface which consists of erosion of land, silting up of river, lakes, accumulation of sand and shingle along the sea coast *etc.*, and also topography and elevation of land from sea level. Strong topographical relief, such as steep hills and deep valleys, has a profound effect on vegetation, chiefly, because it produces characteristic "local climates".

☆ Topography is the nature of surface earth (leveled or sloppy) is known as topography. Topographic factors affect the crop growth indirectly. Altitude

– increase in altitude cause a decrease in temperature and increase in precipitation and wind velocity (hills and plains). Steepness of slope: it results in run off of rain water and loss of nutrient rich top soil. Exposure to light and wind: a mountain slope exposed to low intensity of light and strong dry winds may results in poor crop yields

E.   Socio-economic Factors

☆ Society inclination to farming and members available for cultivation. Appropriate choice of crops by human beings to satisfy the food and fodder requirement of farm household.

☆ Breeding varieties by human invention for increased yield or pest and disease resistance. The economic condition of the farmers greatly decides the input/resource mobilizing ability.

Chapter 14

Plant Ideotypes, Adaptations and Distribution of Crops

Ideotype

☆ Refers to plant type in which morphological and physiological characteristics are ideally suited to achieve high production potential and yield reliability.

☆ The concept of ideotype was given by Donald in 1968. He illustrated that there should be minimum competition between the crops and crop must be competent to compete with weeds. The single plant would give the better result in a group when the crop has at least competition with the same type of the crop.

☆ Ideotype is the model type which may also be defined as "a biological model which is expected to perform or behave in a predictable manner within a defined environment". On the basis of environment Donald and Hamblin (1976) identified two forms of ideotypes *i.e.* isolation ideotypes and competition ideotypes. Competition ideotypes are suitable for mixed cultivation.

☆ Plant breeders have developed an impressive range of techniques in their search for increased yield and better quality in crops. Mutation breeding, polyploidy, the exploitation of hybrid vigour, embryo culture and advanced statistical design and analysis are among the many procedures which have enabled more effective breeding programmes.

Definition

In broad sense an Ideotype model which is expected to perform or behave in a predictable manner within a defined environment. More specifically, crop Ideotype is a plant model which is expected to yield greater quantity of grains, fibre, oil or other useful product when developed as a cultivar. The term Ideotype was first proposed by Donald in 1968 working on wheat.

Characteristics Features of Ideotypes

☆ Crop Ideotype refers to model plants or ideas plant type for a specific environment.

☆ Ideotype differs from Ideotype. The former refers to a combination of various plant characters which enhance the yield of economic produce, whereas the latter refers to the morphological features of the chromosomes of a particular plant species.

☆ Donald included only morphological characters to define an Ideotype of wheat, subsequently, physiological and biochemical traits were also included for broadening the concept of crop Ideotype.

☆ Ideal plants or model plants are expected to give higher yield than old cultivars in a defined environment.

☆ Ideotype is a moving goal which changes according to climatic situation, type of cultivation, national polley, market requirement *etc.* In other words, Ideotype have to be redesigned depending upon above factors. Thus, development of crop Ideotype is a continuous process.

☆ Ideal plant type or model plant type also varies from species. Moreover, this is a difficult and slow method of cultivar development because various morphological, physiological and biochemical characters have to be combined a single genotype from different sources.

☆ Donald (1968) proposed the ideotype approach to plant breeding in contrast to the empirical breeding approach of defect elimination and selection for yield per se. He defined "crop ideotype" as an idealized plant type with a specific combination of characteristics favorable for photosynthesis, growth, and grain production based on knowledge of plant and crop physiology and morphology.

Types of Ideotype

☆ *Isolation ideotype:* It is the model plant types that perform best when the plants are space-planted.

☆ *Competition ideotype:* This ideotype perform genetically well in heterogeneous population. In case of cereals, this ideotype is tall, leafy, free-tillering plant that is able to shade its aggressive neighbours. In case of annual seed crops, such an ideotype will include the following features:

annual habit, tallness, leafy Canopy, tillering or branching,seed size,speed of germination and root characters

☆ *Crop ideotype:* This ideotype perform best at commercial crop densities because it is poor competitor. In case of cereals, a crop ideotype is erect, sparsely-tillered plan, with small erect leaves.

Other Ideotypes are:

☆ *Market ideotype:* includes traits like seed colour, seed size, cooking and baking quality, *etc.*

☆ *Climatic ideotype:* includes trait heat and colds important in climate adaptation such as heat and cold resistance, maturity duration, drought resistance *etc.*

☆ *Edaphic ideotype:* includes salinity tolerance, mineral toxicity/deficiency tolerance *etc.*

☆ *Stess ideotype:* shows resistance to both biotic and abiotic stress.

Adaptation and Distribution of Crop

Distribution

☆ Distribution of commercial crop production throughout the world is governed by many factors, with the main ones being climate, soils, topography, insect pressure, plant disease and economic conditions.

☆ Crop adaptation is determined primarily by genotype-environment interaction, the suitability of a crop to a particular region depending largely on the climatic features of the region in relation to the requirements for normal growth and development of the crop. Accepting climate as

☆ A dominant factor governing crop distribution.

Adaptation

☆ Adaptation of crop plants depends on many factors, and is best considered in relation to a set of conditions (environmental, edaphic (soil) and biotic) rather than to a single factor alone. In many situations, one factor (*e.g.* water availability) may dominate the prevailing conditions, and the nature of the plant's response then largely reflects its adaptation to the existing level of that factor.

☆ More typically, adaptation is expressed as a response to a combination of factors (*e.g.* temperature and daylength) and the nature of the response then reflects the plant's adaptation to the factors in combination.

☆ *Success of a plant in a particular environment* rarely depends on possession of a single adaptive character. Rather, fitness or adaptation to an environment depends on possession of an optimum combination of characters that minimises the deleterious effects and maximises the advantageous effects. The task of plant breeders is thus difficult and complex, as they generally have to develop genotypes with an optimum combination of adaptive characters, rather than ones with a single adaptive character. Whatever the growing conditions, the important consideration is the nature of the adaptive plant response itself and, for commercial purposes, the consequences of that response in terms of the economic output of the crop.

The concept of adaptation can be difficult to define, as it is used in respect to both the evolutionary origins of a character and its contribution to the fitness of the plant to survive in its present environment. Adaptation is also heritable, i.e. it is determined by the genotype of the plant. Hence the definition can be refined to 'the heritable modifications to a plant which enable it to survive, reproduce, or both, in a given environment' (Kramer, 1980). Reproduction, as well as survival, is a critical consideration in the commercial production of seed (grain) crops, as their economic product results from successful completion of the reproductive phase of their life cycles. In these crops, completion of all phases of development is fundamental to economic performance. In other crops such as sugar cane or forages, where the economic product is biomass (plant dry matter) that results from vegetative growth, development is a less important consideration than growth.

☆ For example, a plant that grows well under a given set of conditions, but fails to flower and set seed, is of little value as a grain crop in that situation. It may, however, be an excellent forage crop under those conditions, as the economic product (leaves and stems) is not dependent on flowering and seed set.

☆ Adaptation was described by Wilsie (1962) thus: 'an adaptation may be defined as any feature of an organism which has survival value under the existing conditions of its habitat. Such a feature or features may allow the plant to make fuller use of the nutrients, water, and temperature or light, available, or may give protection against adverse factors such as temperature extremes, water stress, disease or insect pressures'.

Crops Distribution in India

Food Crops

☆ Food crops cover most of the total cropped area in the country and contribute to about 50 per cent of the total value of agricultural production.

☆ They are grown throughout the country either as a sole crop or in combination with other crops.

Rice

- ☆ Rice (*Oryza sativa*) is the leading crop of India and its growth area stretches from 8° N latitude to 34° N latitude. Rice is also grown in areas below sea level as in the Kuttanad region of Kerala.

- ☆ In Andhra Pradesh the deltas of Krishna and Godavari and the adjoining coastal plains form one of the most important rice tracts in the country. Rice is grown both in kharif and rabi seasons. The districts of the East and West Godavari, Kurnool, Anantpur, Krishna, Srikakulam, Visakhapatnam, Nellore and Cuddapah are the main places where rice is largely raised.

- ☆ In Assam rice is the main food crop. It is raised in the Brahmaputra valley including Goalpara, Kamrup, Darrang, Lakhimpur, Sibsagar and Nowgong districts and the Barak valley in Cachar district. A substantial amount of the crop is produced under shifting cultivation system, locally called jhum. In Assam the winter crop is the most important followed by the autumn and the summer crop.

- ☆ In Bihar the main regions of rice cultivation are Shahabad, Champaran, Gaya, Darbhanga and Purnia, while Santhal Parganas, Ranchi and Singhbhum are the main rice producing centres of Jharkhand. In Bihar and Jharkhand, autumn rice is sown in May-June and harvested in September. But winter rice is sown during May-June, transplanted in June-July and harvested in October-November.

- ☆ Over 90 per cent of Orissa's rice comes from Sambalpur, Dhenkanal, Cuttack, Puri, Balasore, Ganjam, Kendrapara, Koraput, Mayurbhanj, Bolangir. Rice occupies about 58 per cent of the state's total cropped area.

- ☆ In Uttar Pradesh, the rice cultivation is confined to Saharanpur, Deoria, Gonda, Bahraich, Basti, Rai Barelli, Lucknow, Varanasi and Gorakhpur. The crop is extensively grown in the eastern and north-eastern parts.

- ☆ In Uttarakhand, rice is grown in Terai region which includes Dehradun also. Rice is also cultivated on the slopes of the lesser and middle Himalayas where it is grown under terracing.

- ☆ In Madhya Pradesh most of the crops are grown in Balaghat, Raigarh and Betul districts, while Raipur, Bilaspur and Surguja districts are main rice growing regions of Chhattisgarh. Rice is largely raised in Tapti, Mahanadi and Narmada valleys.

- ☆ In Tamil Nadu, north Arcot and Thanjavur districts in the Cauvery delta account for 60 per cent of the state's production. Chingleput, Tirunelveli, Tiruchirapalli and Ramanathapuram are other leading districts.

- ☆ Rice in West Bengal accounts for more than 60 per cent of the sown area in every district. The winter crop (aman paddy) is the most important accounting for over two-thirds of the state's production followed by the autumn crop (aus paddy). However, there has been greater emphasis on the cultivation of high yielding varieties under the summer crop (boro paddy) particularly in the irrigated tracts. Cooch Behar, Jalpaiguri, Bankura, Midnapore, Dinajpur, Burdwan and Darjeeling are important districts for rice production.

☆ Punjab has become a major rice producing state with the help of irrigation. Patiala, Jalandhar, Amritsar and Faridkot districts are the areas for rice cultivation.

Wheat

☆ Next to rice, wheat (*Triticum*) is the most important food crop. The germ wheat (*Triticum*) has several species, *viz., Triticum durum, Triticum aestiyum* L., *Triticum compactum, Triticum spelta, Triticum dicoccum, etc.* However, in India the common bread wheat varieties are *Triticum aestiyum* L., the macaroni wheat (*Triticum durum*) and Emmer wheat (*Triticum dicoccum*). *Triticum dicoccum* is grown on a very restricted scale in Gujarat, Maharashtra, Andhra Pradesh and Tamil Nadu where it is known under the names of popaliya, khapli, rava, godhumalu and samba respectively.

☆ *Triticum durum* is the second most important wheat specie grown in the country. This specie is grown mostly under rainfed conditions in Madhya Pradesh, parts of Gujarat and Rajasthan, Maharashtra and Karnataka. Only recently, with the development of dwarf high yielding varieties, some area has come under dwarf durum in Punjab, central and peninsular India. Good quality pasta wheat suitable for macaroni, spaghetti, vermicelli and noodles are now available.

Based on the agro-climatic conditions, the India is broadly divided into five wheat zones:

I. The North-Western Plains consisting of the plains of Punjab, Haryana, Jammu, Rajasthan and western Uttar Pradesh. Here the irrigated wheat is planted in November and the rainfed wheat towards the end of October. Harvesting generally starts by the middle of April and goes up to the beginning of May. This zone is most important among the five zones and *Triticum aestivum* is mostly grown here.

II. The North-Eastern Plains zone consisting of eastern Uttar Pradesh, Bihar, West Bengal, Assam, Odisha, Manipur, Tripura, Nagaland, Meghalaya, Mizoram, Arunachal Pradesh and Sikkim. Here also *Triticum aestivum* is grown. Due to late harvesting of paddy, most of the sowing of wheat is generally done towards the latter half of November and in the first fortnight of December. Harvesting is done in March-April.

III. The Central zone consisting of Madhya Pradesh, Rajasthan and Bundelkhand area of Uttar Pradesh. Both *T. aestivum* and *T. durum* are grown in this zone.

IV. The Peninsular zone consists of the southern states of Maharashtra, Andhra Pradesh, Karnataka and Tamil Nadu. All the three species, namely, aestivum, durum and dicoccum are gown in this zone.

V. The Northern Hill zone consists of the hilly areas of Kashmir, Himachal Pradesh, Uttar Pradesh, West Bengal, Assam and Sikkim. In this zone, the sowing is done in October and the harvesting is done in May/June.

About 80 per cent of the country's total wheat output comes from Uttar Pradesh, Punjab, Haryana, Madhya Pradesh, Rajasthan and Bihar while the remaining comes from West Bengal, Gujarat, Himachal Pradesh, Maharashtra, Jammu and Kashmir, and Karnataka. Uttar Pradesh and Uttarakhand contribute about 35 per cent of total production and are the leading wheat-growing states. Wheat lands are concentrated mainly in the Doabs between the Ganga and the Ghagra Rivers and between the Ganga and the Yamuna. The largest wheat- producing district is Gorakhpur followed by Meerut, Bulandshahr, Etawah, Moradabad, Shahjahanpur, Nainital, Jhansi, Hamirpur and Banda. Thus the crop is concentrated in the north-western and mid-western districts where rainfall and irrigation facilities are better.

Increasing trend in the wheat production in India is mainly attributed to Punjab. Wheat is an important crop in almost every district—Jalandhar, Ludhiana, Faridkot, Bhatinda, Patiala, Gurdaspur, Amritsar, Sangrur, *etc.* The main factors for its production are abundance of fertile alluvium, large irrigational facilities and higher yield per hectare. Haryana is another important producer of wheat. The hectare yield in Madhya Pradesh is low mainly due to lack of irrigational facilities. The area west of the line joining Katni, Jabalpur and Nagpur raises most of the crop. Paucity of rainfall and irrigational facilities restrict wheat cultivation in Rajasthan where Shriganganagar, Kota, AJ.war, Tonk, Sawai Madhopur, Bharatpur, Jaipur, Chittorgarh, Udaipur and Pali districts are the leading districts for production of wheat. West Bengal has registered an increase both in area and production of wheat. Most of the production comes from Birbhum, Burdwan, Murshidabad districts. Gujarat and Bihar have also registered an increase in production. The important districts for production in Gujarat are Mehsana, Rajkot, Kheda, Sabarkanta, Junagadh and in Bihar, we have Champarcm, Gaya, Monghyr, Patna, Saharsa, Muzaffarpur and Shahabad districts.

Maize

☆ Maize is a foodstuff as well as a raw material for starch, glucose, dextrose, sorbitol, germ oil, fibre and gluten products with application in industries such as alcohol, textiles, paper, cosmetics, and pharmaceuticals. In India maize occupies third place among cereals after wheat and rice. It was introduced by the Portuguese in the 17th century. Maize is grown extensively in regions of humid subtropical climate. It may grow even in desert climate provided there is irrigation.

☆ There are three distinct seasons for cultivation of maize in India: main season is kharif, but its cultivation is done during rabi in peninsular India, Jharkhand and Bihar, and in spring in northern India. Higher yields have been recorded in the rabi and spring crops.

☆ Sowing in rows is generally done with drill or by dropping the seed behind the plough. The practice of broadcasting, particularly under rainfed conditions and for fodder maize, is still prevalent in several parts of the country.

☆ Maize is largely grown in the upper Ganga valley, north-east Punjab and south-western Kashmir and southern Rajasthan. Sixty per cent of the total production of maize in the country comes from the states of Uttar Pradesh, Madhya Pradesh, Chhattisgarh, Rajasthan and Bihar. Punjab, Jammu and Kashmir, Gujarat, Himachal Pradesh and Andhra Pradesh also grow maize.

Jowar

☆ Jowar or great millet (also called sorghum) is the most important food and fodder crop of dryland agriculture.

☆ The leading jowar producing states are Madhya Pradesh, Chhattisgarh, Maharashtra, Karnataka and Andhra Pradesh, accounting for 75 per cent of the total area and 80 per cent of the total production in the country. Remaining production comes from Tamil Nadu, Uttar Pradesh, Gujarat and Rajasthan.

Bajra

☆ This crop is cultivated for grain as well as for fodder in India. It is mainly used as a staple food in north-western Rajasthan and Gujarat.

☆ Bajra is grown under warm and dry climatic conditions. It is grown mostly during June to October, as a winter crop from November to February or as a summer crop from March to June. It is suited to areas of low rainfall. It is seldom grown in areas where rainfall exceeds 100 cm. The ideal temperature for its growth is between 25 °C and 35 °C.

☆ It is generally grown on a wide range of soils such as sandy loams of Punjab and Uttar Pradesh and the light soils of Rajasthan and northern Gujarat, heavy clays of Andhra Pradesh, Tamil Nadu and very light soils as in Marathwada and the shallow black, red and light soils of the Deccan and southern India. However, it is suited to light soils.

☆ The crop is grown either as a pure or a mixed crop. As a mixed crop it is grown with cotton, jowar or ragi. It fits into an intensive cropping pattern of three or four crops per year. The preparation of land is done on a very limited scale, since the traditional areas of cultivation are of light texture.

☆ Rajasthan, Maharashtra, Gujarat, Uttar Pradesh, Haryana and Andhra Pradesh are important bajra-growing states.

Ragi

☆ An important cereal in Karnataka, ragi is used by millions as a staple food. It is extensively grown in Karnataka, Tamil Nadu, Andhra Pradesh, Orissa, Bihar, Gujarat, and Maharashtra and in the hilly regions of Uttaranchal and Himachal Pradesh.

☆ Ragi is grown in areas with rainfall ranging from 50 to 100 cm and in irrigated areas. It is also raised as a summer crop and as a rabi crop in southern India, but mostly during kharif in northern India. The soils favourable to this crop are red loams, black and sandy loams in the south and alluvial soils in Gujarat, Uttar Pradesh, Bihar and Jharkhand.

☆ The irrigated crop is raised throughout the year in Karnataka, Tamil Nadu and Andhra Pradesh. The crop flowers in 60-80 days and matures in about 135 days depending on the tract and the variety. The seeds are sown broadcast or with the help of drills and even transplanted on well prepared friable beds ploughed several times.

Barley

☆ Barley is an important rabi cereal in many parts of northern India. It is of minor importance in the south although it can be grown successfully wherever wheat can be grown. It is used as a bread grain and feedstuff for animals.

☆ Barley does not thrive well in regions of high humidity and high temperatures. It is ideally suited to areas where rainfall is either low, even uncertain. A rainfall of 75 cm per year is good for the plant. Barley does best in areas where the winter is cool and the growing period lasts about five months. Areas that are always warm and moist are not suitable for this crop.

☆ Barley is generally grown on light soils, although well-drained medium loams of moderate fertility and texture are most suitable. It is grown widely on a variety of soils, ranging in their texture from sandy to heavy loams in the Indo-Gangetic plains and on terraced slopes in the hills. It is more tolerant to alkali and saline conditions than other rabi cereals.

☆ For raising a crop under dryland farming, soil and water conservation measures such as deep ploughing and dusking after each rainfall, levelling and bunding should be done for crop growth.

☆ Barley is sown as broadcast or with the help of a drill. The best depth for sowing is three-five cm under irrigation and five-eight cm under rainfed conditions, depending upon the initial soil moisture.

☆ Barley is mainly grown in northern regions, *viz.*, Haryana, Punjab, Rajasthan, and Himachal Pradesh and also in Jharkhand. Madhya Pradesh, Chhattisgarh and Jammu and Kashmir also contribute to barley production.

Pulses

Pulses are an important source of protein in vegetarian diets. Being leguminous plants, they also restore soil fertility by fixing atmospheric nitrogen. Pulses are grown on marginal rainfed lands and there are few high- yielding varieties.

Gram

☆ Gram (Cicer arietinum) is the most important pulse crop. It is generally a rabi crop in the unirrigated tracts of the Great Plains. The main contributors are Punjab, Haryana, Uttar Pradesh, Rajasthan, Madhya Pradesh, West Bengal and Maharashtra. Gram is generally grown as a dry crop in the rabi season. Sometimes it is also grown as a regularly or partially irrigated crop. It is best suited to areas having low to moderate rainfall and a mild cold weather. Excessive rain soon after sowing or at flowering does great harm.

☆ Gram is grown alone or mixed with wheat, barley, linseed, safflower or mustard. The preparation of land is the same as for wheat, except that no fine tilth is attempted and the soil is not compacted but is left somewhat cloddy. The crop is rarely manured but the application of phosphatic fertilisers has been shown to increase the grain yield.

Tur

☆ Tur (*Cajanus cajan*) or arhar is the most important pulse crop next to gram. It is grown as a dry crop mixed with cereals like jowar, bajra and ragi.

☆ Tur can be cultivated both in hot-moist and dry climates. It is chiefly sown in the kharif season soon after the rains in June-July. Healthy sunny weather during the flowering and ripening stage is needed for copious setting of fruit.

☆ It is largely grown in Uttar Pradesh, Maharashtra, Madhya Pradesh, Karnataka, Andhra Pradesh, Orissa, Bihar, Tamil Nadu and Gujarat.

Black Gram

☆ Black Gram (*Phaseolus mungo*), green gram (*Phaseolus aureus*) and lentil (*Lens culinaris*) are also important pulse crops. The first two are kharif crops grown as subsidiary to cotton, jowar, maize and millets.

☆ In some parts of peninsular India they are grown to replenish soil fertility after sugarcane or rice crops. Lentil is a rabi crop often grown mixed with barley or mustard. The pulses can be grown in light loams, alluvials, black or red soils, and are usually cultivated under rainfed conditions.

Commercial (Cash) Crops

☆ These crops comprise the crops other than food grains, which bring cash to the cultivator. They are produced mostly for sale either in the raw form or in a semi- processed form.

☆ Cash crops grown in India include sugarcane, tobacco, fibre crops and oilseeds. Indian agro-climatic conditions are highly favourable for the production of sugarcane which provides raw material for industries like sugar, paper and alcohol, substitute for petroleum products, and a host of other chemicals.

Sugarcane

☆ Sugarcane (*Saccharum officinarum*) holds an enviable position amongst all the commercial crops. India among all countries has the largest area under sugarcane.

☆ Sugarcane being a tropical crop grows best in areas with temperatures between 20 °C and 28 °C. A long rainy season of eight months' duration in summer with about 150 cm rainfall and short, cool, dry winter season during ripening and harvesting are ideal. For ripening it needs a cool, dry season; but where rainfall is too heavy and prolonged, the quality of the juice tends to be low, and where the weather remains comparatively warm and moist throughout the year, it does not ripen well.

☆ Sugarcane is planted either in furrows or trenches. In northern India planting is usually done with the onset of the warm weather and is completed well before the onset of summer. Thus the first fortnight of March is the best time for planting sugarcane in Punjab and Haryana, February in Uttar Pradesh and January-February in Bihar. In Maharashtra and parts of Karnataka, it is done in December-February for the 12-month crop, in October-November for the 15 to 16-month crop, and in July-August for the 18-month crop.

☆ In Uttar Pradesh, Saharanpur, Bulandshahar, Shahjahanpur, Meerut, Aligarh, Azamgarh, Faizabad, Ballia, Moradabad, Jaunpur and Varanasi are important producing districts where the soil is fertile and irrigation facilities ample.

☆ In Punjab it is raised in Amritsar, Jalandhar, Ferozepur, Gurdaspur, Sangrur, Patiala and Ludhiana districts.

☆ In Haryana, the important producers are Karnal, Ambala, Rohtak, Hissar and Gurgaon districts.

☆ Bihar raises sugarcane in Champaran, Saran, Darbhanga, Muzzaffarpur, Gaya, Bhagalpur, Patna, and Purnia districts.

☆ Andhra Pradesh raises sugarcane in the coastal districts of Srikakulam, Visakhapatnam, East and West Godavari and Krishna and in southern districts of Nizamabad.

☆ In Tamil Nadu, cane is cultivated under irrigation in the Cauvery basin and Mettur dam areas. North and South Arcot, Ramanathapuram, Madurai, Coimbatore, Tiruchirapalli are important producing districts.

☆ In Maharashtra it is largely grown in black soil under irrigation. Kolhapur, Nasik, Pune, Ahmednagar, Satara, Sangli, and Sholapur districts are important producers.

☆ In Karnataka it is mainly raised in the areas of Krishnarajasagar dam, upper Cauvery and Tunghabadra dam. Shimoga, Raichur, Mandhya, Kolar, Bellary and western Belgaum districts are important producers.

Cotton

☆ Cotton (*Gossypium*) India grows on a commercial scale varieties falling under all the four cultivated specie of Gossypium—*G. hirsutum, G. barbadense, G. arboreum* and *G. herbaceum.* The predominant species cultivated is *G. hirsutum.* On the basis of length of fibre cotton is classified into short, medium and long. If the length of the fibre is less than 1 inch it is called short staple cotton. If the length of fibre is more than 11\8 inches it is termed as long staple.

☆ Being a tropical and subtropical crop, the minimum temperature required for successful germination of seeds is 15 °C. The optimum temperature falls below 21 °C. Cotton plant needs sufficient rainfall in the early stages of growth, but a sunny and dry weather after flowering. Cotton is essentially grown as a kharif crop in major parts of the country.

☆ In most areas the irrigated crop is sown from March-May and the rainfed crop in June-July with the commencement of the monsoon. Cotton is grown on a variety of soils. It requires a soil amenable to good drainage, as it does not tolerate waterlogging. Black and medium black soils are considered ideal for cotton.

☆ It is grown as an irrigated crop in the alluvial soils.

☆ The pattern of production has changed over the years with about 40 per cent of production coming from extra long staple cotton, with short staple variety declining.

☆ Gujarat produces most of the cotton. Ahmedabad, Mehsana, Bharuch, Kheda, Vadodara, Sabarkanta, Surat, Panchmahals and Amreli are the main producing districts.

☆ In Maharashtra the principal cotton growing districts are Buldhana, Akola, Yavatmal, Amravati, Wardha, Aurangabad, Nanded, Dhul, Jalgaon, Nagpur, Prabhani and Beed.

☆ In Madhya Pradesh and Chhattisgarh, Indore, Ujjain, Ratlam, Bhopal, Raipur, Dewas, Rajgarh are the main producers.

☆ The cotton growing areas in Tamil Nadu are Coimbatore, Salem, Madurai, Tiruchirapalli, Ramanathapuram, Tirunelveli, South Arcot and Chingelput districts.

☆ The principal producing areas in Karnataka are the districts of Bellary, Hassan, Bijapur, Gulbarga, Mysore, Shimoga, Dharwar, Raichur, Chitradurg, and Chikmagalur.

☆ In Andhra Pradesh, Kurnool, Cuddapah, Adilabad, Anantpur, Guntur and Hyderabad are important districts for cotton.

☆ The main cotton growing districts in Punjab are Patiala, Ludhiana, Gurdaspur, Sangrur, Ferozepur, Hoshiarpur, and Bhatinda. In Haryana, Gurgaon, Karnal, Rohtak, Ambala and Hissar grow cotton.

☆ Saharanpur, Muzaffarnagar, Meerut, Bijnor, Moradabad, Aligarh, Bulandshahar, Mathura, Agra, Etawah, Mainpuri and Rampur districts of Uttar Pradesh grow cotton.

Jute and Mesta Jute

☆ *Corchorus capsularis,* a bast fibre, is obtained from one of the most important cash crops of eastern India. The crop is grown in seven states namely West Bengal, Bihar, Assam, Odisha, Uttar Pradesh, Tripura and Meghalaya. However the principal jute growing districts are in West Bengal, Bihar, Assam and Odisha.

☆ They share about 80 per cent of the total raw jute production of which West Bengal alone accounts for over 50 per cent. Fibre is obtained from the stem of cultivated varieties of two species commonly known as white and tossa. Tossa fibre is generally finer, softer, stronger, more lustrous and with less root content than white fibre. Besides these two species, mesta is regarded as a substitute fibre for jute.

☆ Jute thrives well in climates of moist heat. The optimum requirements are loamy alluvial soil, average relative humidity between 70 and 90 per cent, well distributed annual rainfall between 115 and 240 cm and temperatures between 27 °C and 30 °C.

☆ Mesta, on the other hand, can be grown on almost any type of soil. However rich loam built up with silt suits it most. It grows well in areas with rainfall between 78 and 100 cm. The advantage with Mesta is that it grows on soils where jute cannot grow. Mesta is also preferred as investment required to grow it is much lower than that for jute.

☆ Jute requires a clean, clod-free field with fine tilth. The land is, therefore, ploughed, cross- ploughed and planked several times. All weeds are thoroughly removed. The sowing of capsularis varieties starts in late

February in low lying areas that retain moisture of the previous flood or monsoon. Sowing in midlands and highlands starts with showers in March or April and continues till early June in the western part of the jute belt.

☆ Dinajpur, Murshidabad, Nadia, Hooghly and 24 Parganas are the leading producing districts in West Bengal. Goalpara, Kamrup and Nowgong districts in the lower Brahmaputra valley produce jute in Assam. Cachar is also an important district for producing jute. Purnia in Bihar and Kheri and Bahraich districts in Uttar Pradesh contribute to jute production.

Oilseeds

Oilseeds which constitute an important group among the commercial crops also play an important role in preparation of useful products like medicines, perfumes, varnishes, lubricants, candles, soaps, *etc.* Oil cake, the residue after oil extraction, is important cattle feed, substantial quantities of which are exported. India is one of the leading oilseed producing countries of the world. The principal oilseeds have been dealt with in this section.

Gujarat leads in production of kharif oilseeds while Uttar Pradesh dominates in rabi oilseed crops:

Groundnut

☆ Groundnut (*Arachis hypogaea*), also termed peanut or monkeynut, was introduced in India during the first half of the 16th century from one of the Pacific islands of China where it had been introduced earlier from either Central or South America. The oil content of the seed varies from 44 to 50 per cent depending on the varieties and agronomic conditions. The edible oils are rich in protein and Vitamins A, B and some members of the B_2 group.

☆ The crop can be grown successfully in places receiving a minimum rainfall of 50 cm to 75 cm. The rainfall should be distributed well during the flowering and pegging of the crop. Sandy loams, loams and well drained black soils which allow adequate root turning are suitable for cultivation of groundnut.

☆ Groundnut is raised mostly as a rainfed kharif crop, being sown from May to June, depending on the monsoon rains. In some areas or where the monsoon is delayed, it is sown as late as August or early September. As an irrigated crop it is grown to a limited extent between January and March and between May and July.

☆ Initially the field is given two ploughings and the soil is pulverised well to obtain a good tilth. The third ploughing may be given just before sowing. The kharif crop is sown with a seed drill or with a suitable planter at a depth of 8-10 cm. Spacings adopted differ from place to place.

☆ Groundnut is extensively grown in peninsular India. Nearly 80 per cent of the groundnut area is concentrated in Gujarat, Tamil Nadu, Maharashtra, Andhra Pradesh, and Karnataka. Other productive areas are Rajasthan, Madhya Pradesh, Uttar Pradesh, Punjab and Odisha.

Rapeseed and Mustard

☆ They are important producers of edible oil and under their names several oilseeds belonging to the cruciferae family are grown in India. They are generally divided into four: brown mustard commonly called rai, sarson (yellow and brown), toria and taramira or tara. In trade, sarson, toria and taramira are known as rapeseed and rai as mustard.

☆ Rapeseed and mustard crops are of tropical as well as temperate zones and require relatively cool temperatures for satisfactory growth. The crops grow well in areas having 25 to 40 cm of rainfall. Sarson and taramira are preferred in low rainfall areas, whereas rai and toria are grown in medium and high rainfall areas. Rai may be grown on all types of soils but toria does best in loam to heavy loams. Sarson is suited to light loam soils and taramira is mostly grown on very light soils.

☆ Nearly 90 per cent of the area and production of rapeseed and mustard is contributed by Uttar Pradesh, Rajasthan, Punjab, Chhattisgarh, Madhya Pradesh and Haryana. The rest comes from Assam, Bihar, West Bengal, Odisha, Gujarat, and Jammu and Kashmir.

Sesame

☆ Known as sesame (*Sesamum indicum*), til and ginpelly, it is a rich source of edible oil with its oil content varying from 46 to 52 per cent. Sesamum oil is used as a cooking medium in south India. It is also used for anointing the body, for manufacturing perfumed oils and for medicinal purposes.

☆ Generally it grows in the plains and at elevations upto 1,200 m. It cannot stand frost, continued heavy rain or prolonged drought. Sesamum requires about 21 °C temperature and a moderate rainfall. The crop requires well drained light loamy soils. In the northern states it is mostly grown as a kharif crop and in the southern states it is generally grown during the rabi season.

☆ The kharif and semi-rabi crops are entirely rainfed, whereas summer crop is grown under irrigation. The yield of the kharif crop is poor, whereas those of the semi-rabi and summer crops are high, as they are grown in rich soils and under better management.

☆ Sesamum is mostly grown in the Satluj-Ganga plain and on the Deccan plateau.

Linseed

⭐ Linseed oil is an excellent drying agent used in manufacturing paints and varnishes, oilcloth, waterproof fabrics and linoleum. It is also used as an edible oil in some areas.

⭐ Linseed is mainly confined to low elevations, in areas with the annual rainfall ranging from 45 to 75 cm. The seed crop does well under moderate cold, but the fibre crop grows best in cool, moist climates. Linseed can be grown on different kinds of soils except the sandy and badly drained heavy clays but does best on clay loams. The crop is grown in the rabi season from September-October to February-March.

⭐ Important production centres are Uttar Pradesh Madhya Pradesh, Maharashtra, Rajasthan, Bihar, Karnataka, West Bengal, and Andhra Pradesh.

Castor

⭐ Castor, indigenous to eastern Africa, is generally grown in India for its oil yielding seeds. The oil content of the seeds varies from 35- 58 per cent in different varieties. The oil is used as a lubricant in the manufacture of soaps, transparent paper, printing inks, varnishes, linoleum and plasticizers. It is also used for medicinal and lighting purposes.

⭐ Castor grows well in relatively dry, warm regions having a well distributed rainfall of 50 to 75 cm. In heavy rainfall areas, the crop puts on excessive vegetative growth and assumes a perennial habit. It also requires a moderately high temperature (20 °C – 26 °C) with low humidity throughout the growing season to produce maximum yields.

⭐ The crop is sown mostly in June-July and to a limited extent in August-September. It is raised chiefly as a rainfed annual crop but sometimes it is planted on the bunds of irrigation channels and borders of garden crops. Andhra Pradesh is the largest producer followed by Gujarat, Orissa, Karnataka and Tamil Nadu.

Sunflower

⭐ Principal states producing sunflower are Karnataka, Maharashtra and Andhra Pradesh. In these states a shift from kharif to rabi culture has taken place owing to the vagaries of monsoon in kharif season and better seed filling in rabi.

⭐ Sunflower is now making inroads into the non-traditional northern states of Punjab, Haryana, Uttarakhand, Uttar Pradesh and Rajasthan as a spring crop.

Soyabean

☆ Among the oilseed crops in India, soyabean occupies third place. Madhya Pradesh leads in area as well as production accounting for about 80 per cent share.

☆ Rajasthan, Maharashtra and Tamil Nadu are other major producers. About 85 per cent of soyabean produced is used for oil, ten per cent for seed and five per cent for food.

Safflower

Safflower is grown in Karnataka, Maharashtra, Andhra Pradesh and Gujarat.

Tobacco

☆ Tobacco (Nicotiana) plays a vital role in the national economy, though it is grown on only 0.25 per cent of the total cropped area. The species Nicotiana tobacum is grown in almost all the states whereas the cultivation of Nicotiana rustica is confined to the northern and north-eastern states, where the temperatures are considerably lower during growing season.

☆ The crop thrives best under moderate temperature and moderate to heavy rainfall conditions. It is highly susceptible to frost. Excessively heavy rainfall however reduces yield and increases the acid content in the leaf. The soils that are best suited are light sandy and sandy loam soils with a low water retentivity capacity.

☆ Such soils produce fine textured leaves which are large, thin and light in colour and body. Heavy soils would produce dark-coloured small leaves of heavy body and strong aroma. The soil must be well drained.

☆ Tobacco being a transplanted crop requires heavy manuring and constant seeding. Experiments have shown that deep ploughing of the fields once in two years during summer followed by 2 to 3 ploughings and 1 to 2 harrowings are beneficial for most of the tobaccos, particularly for the flue cured Virginia tobacco grown in Andhra Pradesh.

☆ Virginia and natu tobacco in Andhra Pradesh are sown in August-September and in Karnataka in April-May; for bidi tobacco in Gujarat and Karnataka, the nurseries are sown in June-July; for the cigar, cheroot and chewing tobaccos in Tamil Nadu, in August-September; for the hookah and chewing tobaccos in Bihar, Uttar Pradesh and West Bengal, in August-October.

Plantation Crops

Tea, coffee, rubber, cardamom, pepper, chilli, turmeric and coconut are the principal plantation crops of India, though they are not equally well organised. Tea is, of course, the gigantic leader. Though pepper and cardamom are native to

India, they are not generally grown, processed or marketed under the accepted plantation system. Therefore, only tea, coffee and rubber are truly regarded as the principal plantation crops of India.

Tea

☆ The name 'tea' (*Camellia thea*) is given to the dried leaves of a broad-leaved evergreen plant known as Thea-Sinensis. There are two main varieties of tea—Chinese and Assamese. Three main types of tea are black, green and brick. Black tea is mainly found in India, Pakistan, Sri Lanka and Indonesia.

☆ The green tea is prepared in China, Japan and Thailand. The main difference between black and green tea is that the former is fermented while the latter is not. Brick tea, on the other hand, constitutes inferior leaves and dust tea, all mixed up and rolled into lumps after mixing with rice paste.

☆ Tea is grown on a plantation scale where the climate' is moist and warm. Rainfall should average between 200 cm and 245 cm per year. The ideal maximum monthly temperature is 24 °C to 30 °C. When the maximum temperature in shade, falls below 24 °C or the minimum temperature below 18 °C-the growth is retarded. Temperature below 10 °C and dry spells affect the crop adversely. The plant requires well drained deep friable loams or forest land rich in organic matter.

☆ The area intended for planting is first cleared of forest growth and adequate measures should be taken to prevent soil erosion. The plants are initially raised in nurseries. Tea is generally propagated from seed but in recent years the use of high yielding clonal material has become popular. The tea estates are highly concentrated in a few hilly districts of India.

☆ The tea in north-east India is grown in an area forming an equilateral triangle, joining Darjeeling in West Bengal, Sadiya in Assam and going beyond Indian borders to Chittagong in Bangladesh. The areas in this region are as follows.

❖ Brahmaputra Valley accounts for about 40 per cent of the total tea area of the country and about 45 per cent of the total production. Maximum concentration of tea cultivation occurs in Lakhimpur, Sibsagar, Tejpur and Bishnath districts. During July to November the maximum crop is harvested.

❖ The Surma valley in Cachar district is another important area contributing about five per cent of the country's total.

❖ The Doars comprising Cooch Behar and Jalpaiguri districts of West Bengal account for 18 per cent of the country's total production.

❖ In Darjeeling district of West Bengal tea gardens are found on hill slopes up to 1,800 m height. Tea is grown in a variety of soils from grey sandy loams to yellow sandy loams, red soil and grey brown forest soils. Maximum area is found over yellow sandy soils which account for about 41 per cent of the tea area of the state giving high yields per hectare.

✰ In South India tea estates are concentrated mainly in the states of Kerala, Tamil Nadu and Karnataka. These together constitute 20 per cent of the country's area and 25 per cent of the country's production. The important tea producing areas are central Thiruvananthapuram, Wynalad, Cochin, Malabar Coast, Nilgiri, Madurai, Kanyakumari and Mysore districts. Tamil Nadu has the highest yield followed by Karnataka at the state level.

Coffee

✰ Coffee (coffea), one of the important plantation crops in India, was started in India in Chickmagalur (Karnataka) in 1826, in Manantody (Wynaad) and Shevaroy in 1830 and the Nilgiris in 1839. It was introduced by a Muslim fakir, Bababudan Sahib, during the 17th century.

✰ Now the plantations are spread over vast hilly tracts of Karnataka, Kerala and Tamil Nadu in the Nilgiris, Cardamom, Palani and Annamalai hills. It is also grown on a limited scale in some non- traditional areas in Orissa, West Bengal and Assam.

✰ Two main species of coffee are generally found in India—arabica (*Coffea arabica*) and robusta (*Coffea canephora*). *Coffea arabica* grows well at elevations between 900 and 1,200 m while *Coffea canephora* grows at lower elevations (about 150 m).

✰ Climatic and environmental factors such as rainfall, temperature and elevation play an important role in determining the conditions of growth of coffee. Warm and humid climate with 180 cm to 200 cm rainfall and 15-30 °C temperature are the main requirements for growth. Its cultivation therefore is restricted to areas where the mean annual temperature does not exceed 28 °C. Soil should be deep, friable, porous, rich in organic matter, slightly acidic.

Rubber

✰ The name 'rubber' (*Hevea brasiliensis*) is given to the latex derived from certain trees. The most important of these trees is *Hevea brasiliensis*. Rubber was first introduced in India in the late 19th century on the banks of River Periyar in north Thiruvananthapuram (Kerala). The seeds of this plant (Hevea) were brought to India from Para (Brazil) by Sir Henry William in 1876. Rubber in India now is mainly grown in the southern parts of India. Kerala, Karnataka and Tamil Nadu are the principal producers. Both in respect of acreage and production Kerala dominates.

☆ The *Hevea brasiliensis* is a hardy, tall, quick growing tree. The main conditions for growth are fairly distributed annual rainfall of not less than 200 cm; a warm humid equable climate of 21 °C to 35 °C; and a well drained deep loamy soil though it can be grown on soils varying from laterite to fine alluvium or clayey loams. Long droughts are injurious to the plant.

☆ Hevea is propagated through seeds and by adopting vegetative methods. The seedlings are usually transplanted in prepared pits in the field in June-July. Rubber trees should be regularly manured with balanced fertiliser mixtures from the time of planting to the stage of economic production for ensuring maximum production.

☆ Latex is obtained from the bark of the rubber tree by tapping. It is cut in spirals to induce the flow of latex. A zinc or iron spout is kept at the base and below the spout a coconut shell is placed to collect the latex. To obtain the optimum yield tapping should be deep.

☆ The forms in which the crop from the rubber plantations are marketed are preserved latex and latex concentrates, dry ribbed sheet rubber, dry crepe rubber, dry solid-block rubber.

☆ The crop collected in the form of liquid latex can be processed in any of the above forms. But the crop collected in the form of tree lace, shell scarp and earth scrap is processed only into crepe or solid block rubbers. The major quantity of natural rubber produced in this country is marketed in the sheet form.

☆ Kerala accounts for most of the rubber plantations. Next in importance are Tamil Nadu, Karnataka, and the Andaman and Nicobar Islands. Kottayam, Kollam, Thiruvananthapuram, Ernakulam, Kozhikode, Cochin and Malabar in Kerala are the chief rubber producers. In Tamil Nadu it is grown in the Nilgiris, Kanyakumari, Coimbatore, Salem and Madurai districts. The important producers in Karnataka are Chikmagalur and Coorg districts. Some natural rubber is also obtained from Assam, West Bengal, Tripura, Meghalaya, Mizoram, Manipur, Nagaland, Orissa, Goa and Maharashtra.

Spices

India is the largest producer of spices with an annual output of two million tonnes. It is also the largest consumer of spices. Amongst the most important spices are pepper, cardamom, chillies, turmeric and ginger.

Pepper

☆ *Piper nigrum* is the earliest known spice crop in India considered to be indigenous to the rain forests in south-west India. The black pepper and the white pepper of commercial use are the dried and processed berries and have a very prominent place in the world market. Pepper is used as a flavouring agent for food stuff and also as a carminative.

☆ Pepper, a tropical crop, *i.e.*, a plant of the humid tropics, grows best in well drained clay loam soil, rich in humus and flourishes in warm, moist climate. An annual rainfall of 250 cm and temperature ranging from 10 to 40 °C is ideal for its cultivation. Red, laterite virgin soil on the slopes of the Western Ghats is also suitable for growing pepper.

☆ The pepper cultivation of India is concentrated in Kerala, Karnataka and Tamil Nadu. Kerala by far is the most important pepper producing state in India, cultivation being concentrated in Kannur, Kottayam, Thiruvananthapuram, Kollam, Kozhikode and Ernakulam districts. North and South Kanara, Coorg, Shimoga, Chickmagalur and Hassan districts of Karnataka, and Kanyakumari and the Nilgiri districts in Tamil Nadu are major producers.

Chilli

☆ Chilli, also called 'red pepper', is an important cash crop in India, which was introduced from Brazil during the 17th century. Capsicum plants are herbaceous or semi woody annuals or perennials.

☆ Moderate rainfall of 60-125 cm and a temperature range of 10-30 °C are the suitable climatic conditions for chilli cultivation. Heavy rainfall and frost affect the crop adversely. The rainfed crop does well on deep, fertile, well drained black cotton soils and somewhat heavy clayey loams. With some irrigation provision, chilli can also be grown on sandy and light alluvial loams and red loamy soils.

☆ Most of the states in India produce chillies to some extent. But the main producers among them are Tamil Nadu, Andhra Pradesh, Rajasthan and Maharashtra.

Cardamom

☆ Cardamom (*Elettaria cardamo- mum*) is considered as the 'Queen of Spices'. It is mainly used for flavouring or medicinal and masticatory purposes. The seeds contain two-eight per cent of a strongly aromatic volatile oil as cardamom consists of dried capsules of the fruit of the same name.

☆ The cardamom plant thrives best in tropical forests at altitudes ranging from 600 to 1,500 m, receiving a well distributed rainfall of over 150 cm and a temperature range of 10-35 °C. It thrives best in the shade provided by the forest trees.

☆ It is highly sensitive to wind and drought and, therefore, areas likely to be affected by these conditions are unsuitable for its growth. The crop is raised chiefly on well drained rich forest loam and red, deep, good textured lateritic soils having plenty of humus or leaf mould.

Ginger

☆ Ginger (*Zingiber officinale*) is mainly grown for its aromatic rhizomes, and used as spice as well as medicine. Ginger is believed to be the native of South-East Asia.

☆ Ginger is a tropical crop requiring high temperatures and enough rainfall (125-250 cm) or irrigation. A certain amount of shade is considered to be favourable for plant growth.

☆ It may be grown alone or mixed with shade-giving plants such as banana, pigeon pea, tree castor and cluster bean (guor). Rich and well drained soil is suitable for it. Sandy or clayey loams and red loams and laterites of the Malabar Coast are ideal soils.

☆ Kerala and Meghalaya are major producers. Himachal Pradesh, Madhya Pradesh, Maharashtra, Karnataka, Uttar Pradesh, Orissa, Rajasthan and West Bengal also produce ginger.

Turmeric

☆ Turmeric (*Curcuma longa*) is the dried rhizome of a herbaceous perennial and a native of India or China. An important condiment and a useful dye it is used in drug and cosmetic industries.

☆ Warm and humid climate is essential for turmeric crop. In regions of heavy rainfall tracts of the west coast, it is grown as a rainfed crop and in other areas it is cultivated under irrigation.

☆ Turmeric thrives in well drained, fertile sandy and clayey, black, red or alluvial loams rich in humus and uniform in texture. Rich loamy soils, having natural drainage and irrigation facilities are the best. It cannot stand water stagnation or alkalinity.

☆ Andhra Pradesh and Tamil Nadu are the main contributors of turmeric. Other major producers are Bihar, Odisha, Maharashtra, Meghalaya and Uttar Pradesh.

Coconut

☆ Coconut (*Cocos nucifera*), a perennial palm, is grown extensively in numerous islands and also in the humid coastal tracts of tropical countries. The coconut palm, known as the Kalpa Vriksha or the tree of heaven, is an important tree providing nuts, timber, fibre, oil and leaves used for various purposes. It is mainly cultivated for the nuts from which two important commercial producers are derived, namely oil and oil cake. Coconut husk is used to make coir or coconut fibre. Its timber and the shells of its nut are used as fuel.

☆ Humid tropical climate is mainly suitable for coconut crop. Among the climatic factors affecting the palm rainfall appears to be the most important. It must range between 100 and 225 cm per annum. The palm can withstand even higher rainfall provided the soil is well drained. Regions with long and dry spells are not suited to its growth.

☆ The optimum mean annual temperature for its best growth and maximum yield is about 27 °C with diurnal variation of 6 °C to 7 °C. Frost and low humidity adversely affect the growth and yield of the palm. Coconut does best on sandy loams along sea coasts and in adjoining river valleys. It also grows on red loams, light grey soils, light black soils and peaty soils.

☆ Coconut palm is a cross pollinated perennial crop. Its saplings are raised in nurseries before they are transplanted to permanent sites about one year later. The planting in the nurseries is done before the outset of the monsoon. The tree begins to yield fruit when it is six-seven years old and from tenth year onwards it gives full yield.

☆ India is a leading coconut producing country in the world. Main producers are Kerala, Tamil Nadu, Karnataka and Andhra Pradesh in that order. Other states producing coconut are Maharashtra, Orissa, and West Bengal.

☆ India became the second largest producer of coconuts after having been in the third place for a long while. Programmes have been undertaken for product diversification to broaden demand and to find new export avenues.

Arecanut

☆ *Areca catechu*, commonly known as arecanut or betelnut, is the product of areca or betelnut palm. It is well known for its consumption for masticatory purpose in India and the Middle East. It originated probably in the Sunda Islands. It is now reported to be grown in only four countries namely India, Bangladesh, Sri Lanka and Malaysia.

☆ Areca is a tall stemmed erect palm, which flourishes in regions of heavy well-distributed annual rainfall of 180 cm or more and high temperatures of 15 °C to 35 °C, in a variety of soils—the laterite soils of the west coast, the red loamy soils of the Mettupalayam region—Tamil Nadu and Kerala— the alluvial soils of Assam and West Bengal and the loam of Orissa. The area, however, should be deep and well-drained without a high water table.

☆ The chief pockets of production are distributed in Kerala, Karnataka and Assam where it is grown extensively and to a smaller extent in Maharashtra, and West Bengal.

Chapter 15

Harvesting and Threshing of Crops

Harvesting

☆ Once the crop is matured or fully ripen, they are cut and gathered (Reaping) which are collectively called as harvesting. Harvesting depends on many factors like season, crop variety, maturity period, *etc.* Over-irrigation, irregular sunlight can prolong ripening of crop which thus delays the harvesting time. Early harvesting causes loss of unripened grains while delayed harvesting leads to shedding off of grains.

☆ Besides this, rodents and even birds eat the grains.

Removal of entire plant of economic parts after maturity from the field is called harvesting. Harvesting is an important arts of techniques of crops production. Delay in harvesting is sometimes deteriorates the quality of the product and the market price decreases accordingly. The crop is harvested early, contains high moisture and more immature grains. The yield will be low due to unfilled grains and this product are subjected to insect pest and disease infection during storage. Harvesting of crops at correct time to get good quality grains and higher yield depend on the practical experience of the farmer.

Therefore regular examination of the crop is necessary as harvesting period approaches. The golden yellow colour is the indication of ripened crops for paddy, rice, and wheat. Manually harvesting is done by using sickles but it is a tedious job as well as time-consuming. In recent times, machines called harvesters are used for harvesting, especially in large-scale farming. Followed by harvesting, threshing of the crop has to be performed. Threshing is the process, in which, the collected grains are separated from the chaff by beating or by the threshing machine. In small-scale farming, chaff and grains are separated from each other by a process called winnowing.

Harvesting Time and Method

Harvesting time and method will vary with the crops as follows:

☆ **CEREALS :** Paddy, Wheat, Maize, Jowar, Bajra *etc.* are the cereals. The cereals crop are harvested when ears are nearly ripe and the plants are dried after yellowing and the seed contain 15 per cent moisture in case of paddy and 12 per cent moisture in case of Wheat, Maize, Jowar, Oat, Bajra *etc.* The wheat is harvested when the straw is golden yellow and brittle. The local aman variety, Jower, Bajra *etc.* are harvested when the plants are completely dried. The maize is harvested at fully mature stage and green cob stage depending on the use. The cereal crops are cut close to the ground with the sickle.

☆ **FIBRE CROPS :** Jute is one of the most important fibre crops of our country. Generally the crops get ready for harvesting at 120 days after sowing. Jute can be harvested at flowering stage, pod stage and pod ripe stage. Both the yield and quality of fibre becomes good if jute is harvested at pod stage. The plants are cut close to the ground with the sickle.

☆ **SUGAR CROPS:** Sugarcane is one of the most important sugar crop of our country. The crop matures within 10 – 12 months after planting. Crops maturity is judged by the following symptoms :

❖ The leaves become yellow and the lower leaves are withering up gradually, leaving progressively fewer leaves at the top.

❖ Plant stops growing.

❖ The internodes of mature cane become short in size and produce metallic sound.

❖ The cane become brittle and break easily at nodes.

❖ The buds swell out at nodes and starts sprouting.

❖ Sucrose content is 20 per cent or more. The crops are cut at ground level preferably after digging down the earthed up ridges.

☆ **TUBER CROPS:** Potato, Ginger, Onion, Turmeric, Elephant foot *etc.* are the tuber crops. They are ready for harvest when haulms starts yellowing and falling on the ground. Potato may be harvested at immature stage in order to get higher prices in the market or to vacant the land for the next crop to be grown. But this potato is totally unsuitable for storage as the skins of immature potato are tender and subjected to rotting. The harvesting is done by spade or country plough.

☆ **Oilseed CROPS**

❖ *Mustard :* Mustard is normally ready for harvest when the pods turn yellow and the plant shed leaves. Delay in harvesting causes shattering of seeds. Harvesting of this crops early in the morning minimizes the shattering loss of seed. The crop is harvested by sickle or uprooting the plants.

❖ *Sesamum :* The crop is harvested when the leaves, stems and capsule begin to turn yellow and lower leaves starts shedding and the flowering is also ceased. Harvesting is delayed for the sake of immature pods results in shattering of mature pods. Delay in harvesting causes the shedding of grains. The crop is harvested by sickle.

❖ *Linseed :* The linseed harvested when the plants turn golden yellow and the crop is dead ripe stage for oilseeds. But the crop is harvested at flowering stage when grown for fibre crops. The harvested by uprooting the plants or cutting them close to the ground by sickle.

❖ *Groundnut :* The yellowing of leaves, shedding of older leaves and the development of the proper colour of testa and a dark tint inside the shell are the prominent symptom of maturity. The bunch type variety is harvested by pulling out the plants when there is adequate moisture in the soil. The spreading variety is harvested by digging out the plant with the help of khurpi, spade or by ploughing the field. The pods are then picked up and dried for a week in sun.

❖ *Sunflower :* The sunflower crop matures in 90–100 days. The crop is harvested when the lower side of the heads turns yellow and some of the bracts dry up. The mature heads are cut with sickle and the seeds are separated by beating the heads with the help of wooden or bamboo sticks.

❖ *Safflower :* The crop matures in about 130–150 days after sowing depending upon the variety. The plants are harvested when they hard enough by uprooting or cutting the plants with sickles.

❖ *Castor :* The crops takes six months to mature. The crops is harvested when one or two capsules in a bunch show sign of drying. The harvested capsules are dried in a sun for 4 – 5 days and thereafter the seed is beaten out with sticks or mallets and winnowed.

☆ **PULSE CROPS:** Gram, Pea, Red gram, Lentil, Green gram, Black gram, Cowpea, Soyabean *etc.* are important pulse crop. Most of the pulse crops are harvested when most of the leaves are dried and shed, the plants are drying and the pods ripe. Harvesting is done by cutting the plants with sickle or uprooting the plants. The pods of some pulse crops such as pea, cowpea *etc.* are picked at periodical intervals, when the pods are well filled and the colour of pods are changing from dark to light green.

The pods of some pulse crops such as green gram, black gram, red gram *etc.* are picked when the pods become mature. These crops after third picking are harvested by cutting the crops with sickle or uprooting the plants. Delay in harvesting causes the shattering of seeds.

✰ **VEGETABLE CROPS**

❖ *Cabbage :* The harvesting time after transplanting depends on the variety. *viz.* :

   ❑ *Early Variety :* 70 – 80 days.

   ❑ *Late Variety :* 100 – 120 days.

   ❑ Cabbage is harvested when the heads are large and firm enough for use. Harvesting may be done by sickle or any other convenient implements. The head is grasped on one hand, bent slightly and cut off with a heavy knife or sickle or hatchet.

❖ *Cauliflower :* The harvesting time after transplanting varies according to the variety *viz.* :

   ❑ *Early variety :* 60 – 70 days.

   ❑ *Mid season variety :* 90 – 100 days.

   ❑ *Late Variety :* 110 – 120 days.

   ❑ Cauliflower should be harvested when the curds reaches a proper size, compact and bright in colour but has not broken into segments. The plant is cut off well below the head and leaves are timed with knife.

❖ *Knolkhol :* Knolkhol should be harvested when the swollen stem reach diameter of 5 to 7 cm. before it becomes tough and woody. The plants are pulled out from the land, it is generally marketed after removing both leaves and roots.

❖ *Brinjal :* Brinjals are harvested when they attains a good size and colour and still they are immature. Mature fruits are not so desirable. Fruits are allowed to attain a good size and colour till they do not lose their bright and glossy appearance. Harvesting is done by detaching the fruits from the stem.

❖ *Tomato :* Tomato are harvested in the following stages according to the uses of the fruits :

   ❑ *Green Stage :* Tomatoes in this stage are harvested for distant places.

   ❑ *Pink Stage :* The fruits in this stage becomes red. But the fruits are not fully ripe. The tomatoes in this stage are harvested for local market.

   ❑ *Ripe Stage :* The surface of most of fruits in this stage is red and the softening of fruits begins. Tomatoes in this stage are harvested for home or table purpose.

❐ *Full Ripe Stage :* Maximum colour development of the fruit at this stage occurred and they feel soft to the touch. The fruits are ordinarily used within 24 hours of picking and are consumed or used for canning and pickling.

☆ ROOT CROPS

❖ **Radish :** Radish becomes ready for harvesting within one month of sowing and they should be harvested according to the variety when the roots are still tender and crisp but they big enough for the market or home consumption. Delay in harvesting makes the root pithy and quite unsuitable for marketing.

❖ **Turnip :** Turnip takes about 70 to 100 days to mature and becomes ready for harvesting. The turnip is harvested when the roots of turnip become 5 to 7.5 cm in diameter. Delay in harvesting makes the root hard and fibrous and their quality is deteriorated rapidly.

❖ **Carrot :** The carrots are harvested when their roots are 2.5 to 3.7 cm in diameter at the upper end. The crops are harvested by pulling out the plants.

❖ **Beet :** Beet becomes ready for harvesting within 2 – 3 months of sowing. The beets are harvested when their roots are 5.0 – 7.5 cm in diameter. The crops is harvested, by pulling out the plant digging with spade.

❖ **Sweet Potato :** Sweet potato becomes ready for harvesting from 105 – 150 days after planting and they are harvested when the leaves turn pale and later slightly yellow and cut surface dries up quickly and does not turn black as that of immature tubers. The vines are removed by cutting and then harvesting of tuber is done with the help of a spade.

☆ LADY'S FINGER :
Flowering begins from 35–40 days after sowing and fruits are ready for harvest four to five days after flowering. The pods may be harvested continuously at some intervals. Delay in harvesting may make the fruits fibrous and they lose their tenderness and taste.

☆ CUCURBITACEOUS VEGETABLES :
Pumpkin, bottle gourd, bitter gourd, snake gourd, ridge gourd, sponge gourd, cucumber, pointed gourd, water melon, muskmelon *etc.* are the important cucurbitaceous vegetables. These vegetables are harvested when the fruits are still young and tender with the exception of pumkin, which is harvested either in green stage or mature stage according to the demand of the market. The water melon and muskmelon should be harvested at proper stage of maturity.

☆ FRUIT CROPS:
Mango, Jackfruit, Banana, Guava, Papaya, Grape, Pineapple, Litchi, Apple, Pear *etc.* are well known fruit crops in our country. Most of the fruits are harvested at ripe and half ripe stage depending on their mode of uses. The full ripe fruits are harvested at full ripe stage if they

are consumed at home. The banana and papaya are harvested at green stage when they are big and firm enough for using as vegetables.

Threshing

☆ Threshing is the method of separating grain from the stalk it grows on and from the chaff or unit that covers it. The edible portion of the crop is loosened in the process, but not the portion of the fibre.

The process of separating the seeds from the ears is known as threshing. In cereals, straws and grains are separated and in pulses the seeds are separated from pods. Threshing in generally done immediately after harvest.

After harvesting and before winnowing, it is done. The method used in old times was to hit with a thrash on the harvested ears of grain and this was performed manually. Instead, goats, donkeys, or bulls squeezed the grain out of the stalks. After this, when the grains were winnowed to remove the debris, the straw was collected and raked away. The air current blew away the lightly weighted waste particles during winnowing, leaving only the hard grain particles.

☆ Andrew Meikle invented the threshing machine later in the seventies. A spinning chamber furnished with wooden mixers was supported by piles of grain. To take away the free straw, the machine had a saw-tooth like a drum and forced the remaining waste and grain into a set of rollers through a strainer that further separated the chaff from the grain before winnowing. In all threshing machines, including the advanced self-moving combinations, the working theory of Meikle's machine was used. The harvesting, threshing, and winnowing are performed by combined harvesters.

☆ In an isolated plot of land called the threshing floor, threshing was normally performed. Some threshing floors were flattened (outdoor) circular or paved surfaces, but the floor used to be a stone or a wooden plank, typically in small-scale farming. Outdoor floors were used by an entire village as a shared land. But, sadly, new devices and technology have outsourced flooring.

Threshing Winnowing

☆ The most important kernel harvester function is threshing. Grain loss and the degradation of the crop contribute significantly to the threshing philosophy and techniques. There are four types of principles of threshing available: scratching, scraping, combing, and grinding.

☆ Grain loss can be considered a peripheral velocity function and affects the pattern of contact. A characteristic of rasp bar contact patterns can be recognized as grain loss. In the subsequent combing threshing process, grain loss resulting from cleaning and separation was significantly decreased.

✩ Winnowing is the name given to the process of separating the grain from the chaff. This is the move that accompanies threshing (the method of loosening the chaff). As the grain is often thicker than the chaff, the gentle wind is normally enough to blow the chaff away, while the grain is left in place. Winnowing often involves ventilation.

Difference between Threshing and Winnowing

Threshing	*Winnowing*
It is a process of beating out grains from stems.	It is a process of separating the grains from the chaff.
It is done by striking harvested crops against a hard surface.	It is with the help of wind separation of grain alone
It is done manually through a machine called thresher or draught animals	This process is carried out by a winnowing machine.
It is done before winnowing	It is done after threshing immediately.

Methods of Threshing

Following are the common method of threshing :

✩ **Threshing by Manual Labour :** It is most common method of threshing and is done mainly by beating against stones or any other hand materials or beating by sticks. It is a slow and labour consuming device, but it is followed for quantity of harvests.

✩ **Threshing by Animals :** It is also very common method of threshing used in village. In this method, the threshing is done by treating the crop by means of a team of animals. A strong pole is fixed in the centre of the threshing space in which bullocks are tied in line one after the other. The harvest is spread on the threshing space and bullocks move round the harvest and trample them continuously till the grains are completely separated from straw. One man drives the bullock from the back. Threshing by treading is also done by tractor. It is most expensive method.

✩ **Threshing by Machines :** It is most improved method of threshing. The threshing machines are used for this purpose. Threshing is followed by winnowing. Winnowing is the method of separating grain or seed from chaff. The seeds or grains are dried in sun to reduce their moisture content. Moisture content for safe storage is 14 per cent for most of the crops. High moisture in grain as well as high humidity in the atmosphere caused sprouting and moulting of grain.

Storage

  ☆ The agricultural produce must be stored properly for the continuous consumption as well as to get the higher price. Because the market value is generally low at harvesting time.

  ☆ Generally the foodgrains, oilseeds, seeds *etc.* are stored for future use.

Factors Affecting Storage

The following factors influences the storage of food grains :

  ☆ *Moisture Content of the Grains :* The seeds that contain high moisture are subjected to the attack of insects and microorganisms. Moist seeds are amenable for easy biting or chewing by insects. Sometimes moist grains may even germinate and become unfit for consumption.

  ☆ *Climatic Condition :* Grains are hygroscopic and absorb moisture from the atmosphere. Under moist climates, the seeds are subjected to the attack of insects and microorganisms as this condition is favourable for the growth and multiplication of insects.

  ☆ *Quality of Produce :* The crops that are harvested early contain more illfilled and shrivelled grains. The shrivelled and broken grains are the predisposing causes for insect attack.

  ☆ *Storage Conditions :* The product is generally stored in gunny bag which is more prone to insect infestation of the grains. The storage condition should be such to provide the grain protection from insects. The ill storage condition results in loses of food grains.

Method of Storage

  ☆ Food grains and oilseeds are stored either in bags or in bulk (*i.e.* without bagging). Pusa bins may also be used for storage of foodgrains.

  ☆ Godowns are the most common structures for above ground bag storage. The godowns should be made free of insects and microorganism by fumigation using EDB (Ethylene-di-bromide).

  ☆ The save grain campaign organisation recommends the uses of EDB ampules as follows:

Quantity of Foodgrains	EDB Ampule	
(Quintal)	Quantity	Number
1 or less	3 ml	1
2	6 ml	1 or 2 of 3 ml.
5	6 ml	3 or 6 of 3 ml.
10	10 ml	3 or 5 of 6 ml.
20	15 ml	4 or 6 of 10 ml.

☆ Several pests attack the produce during storage which can be controlled by spraying malathion and dichlorovas. Fumigants such as Aluminium phosphide, Methyl bromide *etc.* are used in godowns. The pest is killed by the poisonous gas released from the fumigants.

Points to Remember for the Storing of Agricultural Produce

☆ The stores should be at least 0.5 km away from the place which are a source of infection like kilns, flour and bone crusting mills, garbage dumping ground, slaughter house and tanneries. It should be constructed, as far as possible, away from dwelling houses.

☆ The stores should be, as far as possible, situated near a transport head or main road.

☆ There should not be any tree near the stores. Otherwise birds will be a nuisance.

☆ The walls of the stores should be made smooth and the cracks and cravices and rat borrows should be filled up completely. The plinth should be kept 0.75 m. above the ground level.

☆ The floor of the stores should be constructed of either cement concrete or stone with a slope of 4 cm. from the wall of its outer edge to prevent rain water from getting inside the godown through the doors.

☆ The ventilators and windows need to be fitted in such a manner that the store is open to air made air light for desired period.

☆ The floor and loose matters of Katcha stores should be scraped annually in April and replastering should be done the store should be rat proof.

☆ The roof of the godown should be water proof.

☆ The debris of the godown should be collected and burnt before storing the agricultural produce.

☆ The old grains from the store should be disposed of or removed and the store should be throughly cleaned before storing the new produce. White washing the walls and cellings of pucca godown should be done.

☆ The store should be disinfected with Malathion 50 EC or Pyrethin (Dilution–1 : 100) @ 3 litres per 100 sq. metres after cleaning the store. The store may be fumigated with Aluminium Phosphite @ 25 tablets (3 gm. each) per 100 cubic metre space in place of malathion spraying.

☆ The bags should be disinfected by dipping 0.5 per cent malathion (Dilution–1 : 50) –50 EC or in boiling water for a minute in case of bag storage.

☆ The completely dried grains (Moisture content ranging from 12–14 per cent) should be stored. The new and old stocks should not be kept in the same godown.

☆ The grains meant for seed should be treated with Malathion 5 per cent 250 gm. per quintals of grains. The existing practise of mixing D.D.T. or B.H.C. should be strictly avoided as this hamper the germination of seed.

☆ The dunnage comprising either timber pallets, timber squares, palm matting or still better, a layer of poly-thene sheet sandwiched between two layers of malting should be used. The floor space between slacks as well as the dunnage should be sprayed with malathion 50 per cent EC to avoid insect infestation.

☆ The bags should be stacked 60 – 70 cms. away from walls with proper alleys around for inspection and other operations (*i.e.* fumigation, spraying *etc.*).

☆ It is essential to inspect the store fortnightly for pest infestation and rodent damage and also for taking necessary measurement to avoid the losses.

Appendices

Important Formulae and Measurements Used in Agriculture

Leaf area index (LAI) $= \dfrac{total\ leaf\ area}{ground\ area}$

Leaf equivalent ratio (LER) $= \dfrac{sole\ crop}{mixed\ crop}$

Harvesting index per cent $= \dfrac{economic\ yield\ (grain)}{biological\ yield\ (grain + straw)}$

Sodium absorption ratio (SAR) $= \dfrac{Na^-}{\sqrt{ca^{+2} + mg^{+2}}}$

Absolute humidity $= \dfrac{weight\ of\ water\ vapour}{volume\ of\ air}$

Cropping intensity per cent $= \dfrac{total\ cropped\ area\ in\ year}{net\ cultivated\ area} \times 100$

Crop rotation intensity per cent $= \dfrac{no.\ of\ crop\ raised\ in\ a\ crop\ rotation}{duration\ of\ crop} \times 100$

$$\text{Cropping index} = \frac{no.\ of\ crops\ grown\ per\ year}{cropped\ area} \times 100$$

$$\text{Crop water use efficiency} = \frac{crop\ yield}{evapotranspiration}$$

$$\text{Pore space or porosity of soil per cent} = 100 - \frac{bulk\ density}{particle\ density} \times 100$$

$$\text{Moisture in soil per cent} = \frac{loss\ in\ weight}{oven\ dry\ weight} \times 100$$

$$\text{Real value of seed} = \frac{pure\ seed\ \% \times germination\ \%}{100}$$

$$\text{Pure live seed} = \frac{purity\ \% \times germination\ \%}{100}$$

$$\text{Particle density of soil} = \frac{weight\ of\ soil\ solids}{volume\ of\ soil\ solids}$$

$$\text{Bulk density of soil} = \frac{weight\ of\ dry\ soil}{volume\ of\ soil(solids + pores)}$$

$$\text{Intelligence Quotient (I.Q.)} = \frac{mental\ age}{chronological\ age} \times 100$$

$$\text{Ginning per cent} = \frac{weight\ of\ lint}{weight\ of\ seed\ cotton} \times 100$$

$$\text{Relative humidity} = \frac{water\ vapour\ present\ in\ the\ air}{water\ vapour\ required\ for\ saturation} \times 100$$

$$\text{Celsius and Fahrenheit relationship} = \frac{C}{5} = \frac{F-32}{9}$$

Important Measurement

Area

100 m^2 =1 are or 0.025 acre

100 are = 1 ha or 2.47 acres

1 acre = 0.405 ha

100 ha = 1 km^2 or 0.386 mile

Length

10 mm = 1 cm or 0.39 inch

100 cm = 1 m or 39.4 inches

100 m = 1 km or 0.62 mile

Weight

1000 mg = 1 g

1000 g = 1 kg or 2.21 pound

1000 kg = 1 tonne

Water flow

1 Cusec = 28.3 lit of water/sec

1 Cusec = 1000 lit/sec

1 ha mm = 1000 lit or 10^3

1 ha cm = 100000 lit or 10^5

1 ha m = 10000000 lit or 10^7.

List of ICAR Institutes

1. ICAR-Central Island Agricultural Research Institute, Port Blair
2. ICAR-Central Arid Zone Research Institute, Jodhpur
3. ICAR-Central Avian Research Institute, Izatnagar
4. ICAR-Central Inland Fisheries Research Institute, Barrackpore
5. ICAR-Central Institute Brackishwater Aquaculture, Chennai
6. ICAR-Central Institute for Research on Buffaloes, Hissar
7. ICAR-Central Institute for Research on Goats, Makhdoom
8. ICAR-Central Institute of Agricultural Engineering, Bhopal
9. ICAR-Central Institute for Arid Horticulture, Bikaner
10. ICAR-Central Institute of Cotton Research, Nagpur
11. ICAR-Central Institute of Fisheries Technology, Cochin
12. ICAR-Central Institute of Freshwater Aquaculture, Bhubneshwar
13. ICAR-Central Institute of Research on Cotton Technology, Mumbai
14. ICAR-Central Institute of Sub Tropical Horticulture, Lucknow
15. ICAR-Central Institute of Temperate Horticulture, Srinagar
16. ICAR-Central Institute on Postharvest Engineering and Technology, Ludhiana
17. ICAR-Central Marine Fisheries Research Institute, Kochi
18. ICAR-Central Plantation Crops Research Institute, Kasargod
19. ICAR-Central Potato Research Institute, Shimla
20. ICAR-Central Research Institute for Jute and Allied Fibres, Barrackpore
21. ICAR-Central Research Institute of Dryland Agriculture, Hyderabad
22. ICAR-National Rice Research Institute, Cuttack
23. ICAR-Central Sheep and Wool Research Institute, Avikanagar, Rajasthan
24. ICAR- Indian Institute of Soil and Water Conservation, Dehradun
25. ICAR-Central Soil Salinity Research Institute, Karnal
26. ICAR-Central Tobacco Research Institute, Rajahmundry
27. ICAR-Central Tuber Crops Research Institute, Trivandrum
28. ICAR-ICAR Research Complex for Eastern Region, Patna
29. ICAR-ICAR Research Complex for NEH Region, Barapani
30. ICAR-Central Coastal Agricultural Research Institute, Ela, Old Goa, Goa
31. ICAR-Indian Agricultural Statistics Research Institute, New Delhi
32. ICAR-Indian Grassland and Fodder Research Institute, Jhansi
33. ICAR-Indian Institute of Agricultural Biotechnology, Ranchi

34. ICAR-Indian Institute of Horticultural Research, Bengaluru
35. ICAR-Indian Institute of Natural Resins and Gums, Ranchi
36. ICAR-Indian Institute of Pulses Research, Kanpur
37. ICAR-Indian Institute of Soil Sciences, Bhopal
38. ICAR-Indian Institute of Spices Research, Calicut
39. ICAR-Indian Institute of Sugarcane Research, Lucknow
40. ICAR-Indian Institute of Vegetable Research, Varanasi
41. ICAR-National Academy of Agricultural Research and Management, Hyderabad
42. ICAR-National Institute of Biotic Stresses Management, Raipur
43. ICAR-National Institue of Abiotic Stress Management, Malegaon, Maharashtra
44. ICAR-National Institute of Animal Nutrition and Physiology, Bengaluru
45. ICAR-National Institute of Research on Jute and Allied Fibre Technology, Kolkata
46. ICAR-National Institute of Veterinary Epidemiology and Disease Informatics, Hebbal, Bengaluru
47. ICAR-Sugarcane Breeding Institute, Coimbatore
48. ICAR-Vivekananda Parvatiya Krishi Anusandhan Sansthan, Almora
49. ICAR-Central Institute for Research on Cattle, Meerut, Uttar Pradesh
50. ICAR-National Institute of High Security Animal Diseases, Bhopal
51. ICAR-Indian Institute of Maize Research,New Delhi
52. ICAR- Central Agroforestry Research Institute, Jhansi
53. ICAR-National Institute of Agricultural Economics and Policy Research, New Delhi
54. ICAR- Indian Institute of Wheat and Barley Research, Karnal
55. ICAR- Indian Institute of Farming Systems Research, Modipuram
56. ICAR- Indian Institute of Millets Research, Hyderabad
57. ICAR- Indian Institute of Oilseeds Research, Hyderabad
58. ICAR- Indian Institute of Oil Palm Research, Pedavegi, West Godawari
59. ICAR- Indian Institute of Water Management, Bhubaneshwar
60. ICAR-Indian Institute of Rice Research, Hyderabad
61. ICAR- Central Institute for Women in Agriculture, Bhubaneshwar
62. ICAR-Central Citrus Research Institute, Nagpur
63. ICAR-Indian Institute of Seed Research, Mau
64. ICAR-Indian Agricultural Research Institute, Post Box No. 48, Hazaribag 825 301, Jharkhand

CENTRAL AGRICULTURAL UNIVERSITIES

1. Central Agicultural University, PO Box 23, Imphal Manipur
2. Rani Laxmi Bai Central Agicultural University, Jhansi, U.P.
3. Dr Rajendra Prasad Central Agicultural University, Pusa, Samastipur, Bihar

DEEMED UNIVERSITIES

1. ICAR-Indian Agricultural Research Institute, New Delhi
2. ICAR-National Dairy Research Institute, Karnal
3. ICAR-Indian Veterinary Research Institute, Izatnagar
4. ICAR-Central Institute on Fisheries Education, Mumbai

UNIVERSITIES WITH AGRICULTURAL FACULTY

Andhra Pradesh

1. Acharya NG Ranga Agricultural University, Guntur
2. Dr. YSRHU (APHU), Venkataramannagudem
3. Sri Venkateswara Veterinary University, Tirupati

Assam

4. Assam Agricultural University, Jorhat

Bihar

5. Bihar Agricultural University, Sabour, Bhagalpur
6. Bihar Animal Sciences University, Patna

Chhattisgarh

7. Indira Gandhi Krishi Viswa Vidhyalaya, Raipur
8. Chhattisgarh Kamdhenu Visvavidyalaya, Durg

Gujarat

9. Sardar Krushinagar Dantiwada Agricultural University, Dantiwada
10. Anand Agricultural University, Anand
11. Navsari Agricultural University, Navsari
12. Junagarh Agricultural University, Junagarh
13. Kamdhenu University, Gandhinagar

Haryana

14. Chaudhary Charan Singh Haryana Agricultural University, Hisar
15. Lala Lajpat Rai University of Veterinary and Animal Sciences, Hisar
16. Haryana State University of Horticultural Sciences, Karnal

Himachal Pradesh

17. Ch. Sarwan Kumar Himachal Pradesh Krishi Viswavidyalaya, Palampur

18. Dr. Yaswant Singh Parmar University of Horticulture and Forestry, Solan

Jharkhand

19. Birsa Agricultural University, Ranchi

Jammu and Kashmir

20. Sher-e-Kashmir University of Agricultural Science and Technology, Srinagar

21. Sher-e-Kashmir University of Agricultural Science and Technology, Jammu

Karnataka

22. University of Agricultural Sciences, Bangalore

23. Karnataka Veterinary, Animal and Fisheries Sciences University, Bidar

24. University of Agricultural Sciences, Raichur

25. University of Agricultural Sciences, Dharwad

26. University of Horticulture Science, Bagalkot

27. University of Agriculture and Horticulture Sciences, Shimoga

Kerala

28. Kerala Agricultural University, Thrissur

29. Kerala University of Fisheries and Ocean Studies, Panangad, Kochi

30. Kerala Veterinary and Animal Sciences University, Pookode, Wayanand, Kerala

Madhya Pradesh

31. Rajmata Vijayaraje Scindia Krishi VishwaVidyalaya, Gwalior

32. Nanaji Deshmukh Pashu ChikitsaVisvaVidyalaya, Jabalpur

33. Jawaharlal Nehru Krishi Viswa Vidyalaya, Jabalpur

Maharashtra

34. Dr. Balaesahib Sawant Kokan KrishiVidyapeeth, Dapoli

35. Maharastra Animal and Fisheries. Sciences University, Nagpur

36. Vasantrao Naik Marathwada Krishi Vidyapeeth, Parbhani

37. Mahatma Phule Krishi Vidyapeeth, Rahuri

38. Dr. Punjabrao Deshmukh KrishiViswaVidyalaya, Akola

Odisha

39. Orissa University of Agricultural and Technology, Bhubaneswar

Punjab

40. Guru Angad Dev Veterinary and Animal Sciences University, Ludhiana
41. Punjab Agricultural University, Ludhiana

Rajasthan

42. Maharana Pratap University of Agriculture and Technology, Udaipur
43. Swami Keshwanand Rajasthan Agricultural University, Bikaner
44. Rajasthan University of Veterinary and Animal Sciences, Bikaner
45. SKN Agriculture University, Jobner
46. Agriculture University, Kota
47. Agriculture University, Jodhpur

Tamil Nadu

48. Tamil Nadu Agricultural University, Coimbatore
49. Tamil Nadu Veterinary and Animal Sciences University, Chennai
50. Tamil Nadu Fisheries University, Nagapattinam

Telangana

51. Sri Konda Laxman Telangana State Horticultural University, Hyderabad
52. Sri PV Narsimha Rao Telangana Veterinary University, Hyderabad
53. Professor Jayashankar Telangana State Agricultural University, Hyderabad

Uttrakhand

54. G.B. Pant University of Agriculture and Technology, Pantnagar
55. VCSG Uttarakhand University of Horticulture and Forestry, Bharsar

Uttar Pradesh

56. Chandra Shekhar Azad University of Agricultural and Technology, Kanpur
57. Narendra Deva University of Agriculture and Technology, Faizabad
58. Sardar Vallabhbhai Patel University of Agriculture and Technology, Meerut
59. U.P. Pt. Deen Dayal Upadhyaya Pashu Chikitsa VigyanVishwaVidhyalaya Evem Go Anusandhan Sansthan, Mathura
60. Banda University of Agricultural and Technology, Banda
61. Sam Higginbottom University of Agriculture, Technology and Sciences, Allahabad

West Bengal

62. Bidhan Chandra Krishi Viswa Vidhyalaya, Mohanpur
63. West Bengal University of Animal and Fishery Sciences, Kolkata
64. Uttar Banga Krishi Viswavidhyalaya, Cooch Behar

AICRPS/NETWORK PROJECTS

1. AICRP on Nematodes, New Delhi
2. AICRP on Maize, New Delhi
3. AICRP Rice, Hyderabad
4. AICRP on Chickpea, Kanpur
5. AICRP on MULLARP, Kanpur
6. AICRP on Pigeon Pea, Kanpur
7. AICRP on Arid Legumes, Kanpur
8. AICRP on Wheat and Barley Improvement Project, Karnal
9. AICRP Sorghum, Hyderabad
10. AICRP on Pearl Millets, Jodhpur
11. AICRP on Small Millets, Bangalore
12. AICRP on Sugarcane, Lucknow
13. AICRP on Cotton, Coimbatore
14. AICRP on Groundnut, Junagarh
15. AICRP on Soybean, Indore
16. AICRP on Rapeseed and Mustard, Bharatpur
17. AICRP on Sunflower, Safflower, Castor, Hyderabad
18. AICRP on Linseed, Kanpur
19. AICRP on Sesame and Niger, Jabalpur
20. AICRP on IPM and Biocontrol, Bangalore
21. AICRP on Honey Bee Research and Training, Hisar
22. AICRP -NSP(Crops), Mau
23. AICRP on Forage Crops, Jhansi
24. AICRP on Fruits, Bangaluru
25. AICRP Arid Zone Fruits, Bikaner
26. AICRP Mushroom, Solan
27. AICRP Vegetables including NSP vegetable, Varanasi
28. AICRP Potato, Shimla
29. AICRP Tuber Crops, Thiruvananthapuram
30. AICRP Palms, Kasaragod
31. AICRP Cashew, Puttur
32. AICRP Spices, Calicut
33. AICRP on Medicinal and Aromatic Plants including Betelvine, Anand
34. AICRP on Floriculture, New Delhi

35. AICRP in Micro Secondary and Pollutant Elements in Soils and Plants, Bhopal

36. lAICRP on Soil Test with Crop Response, Bhopal

37. AICRP on Long Term Fertilizer Experiments, Bhopal

38. AICRP on Salt Affected Soils and Use of Saline Water in Agriculture, Karnal

39. AICRP on Water Management Research, Bhubaneshwar

40. AICRP on Ground Water Utilisation, Bhubaneshwar

41. AICRP Dryland Agriculture, Hyderabad

42. AICRP on Agrometeorology, Hyderabad including Network on Impact adaptation and Vulnerability of Indian Agri. to Climate Change

43. AICRP Integrated Farming System Research, Modipuram including Network Organic Farming

44. AICRP Weed Control, Jabalpur

45. AICRP on Agroforestry, Jhansi

46. AICRP on Farm Implements and Machinery, Bhopal

47. All India Coordinated Research Project on Ergonomics and Safety in Agriculture

48. AICRP on Energy in Agriculture and Agro Based Indus.,Bhopal

49. AICRP on Utilization of Animal Energy (UAE), Bhopal

50. AICRP on Plasticulture Engineering and Technologies, Ludhiana

51. AICRP on PHT, Ludhiana

52. AICRP on Goat Improvement, Mathura

53. AICRP- Improvement of Feed Sources and Nutrient Utilisation for raising animal production, Bangalore

54. AICRP on Cattle Research, Meerut

55. AICRP on Poultry, Hyderabad

56. AICRP-Pig, Izzatnagar

57. AICRP Foot and Mouth Disease, Mukteshwar

58. AICRP ADMAS, Bangalore

59. AICRP on Home Science, Bhubaneshwar

Network Projects - 19

1. All India Network Project on Pesticides Residues, New Delhi

2. All India Network Project on Underutilised Crops, New Delhi

3. All India Network Project on Tobacco, Rajahmundry

4. All India Network Project on Soil Arthropod Pests, Durgapura

5. Network on Agricultural Acarology, Bangalore

6. Network on Economic Ornithology, Hyderabad
7. All India Network Project on Rodent Control, Jodhpur
8. All India Network Project on Jute and Allied Fibres, Barrackpore
9. Network project on Improvement of Onion and Garlic, Pune
10. Network Biofertilizers, Bhopal
11. Network Project on Harvest and Postharvest and Value Addition to Natural Resins and Gums, Ranchi
12. Network project on Animal Genetic Resources, Karnal
13. Network Project on R and D Support for Process Upgradation of Indigenous Milk products for industrial application Karnal
14. Network Programme on Sheep Improvement, Avikanagar
15. Network Project on Buffaloes Improvement, Hisar
16. Network on Gastro Intestinal Parasitism, Izatnagar
17. Network on Haemorrhagic Septicaemia, Izatnagar
18. Network Programme Blue Tongue Disease, Izatnagar
19. Network Project on Conservation of Lac Insect Genetic Resources, Ranchi
20. Network Project on Agricultural Bioinformatics and Computational Biology, New Delhi

OTHER PROJECTS - 10

1. Technology Mission on Cotton (CICR, Nagpur)
2. Technology Mission on Jute (CRIJAF, Barrackpore)
3. Continuation, Strengthening and Establishment of Krishi Vigyan Kendras
4. Strengthening and Development of Higher Agricultural Education in India, New Delhi
5. Central Agricultutral University, Imphal
6. Strengthening and Modernization of ICAR Headquarters
7. Intellectual Property Management and Transfer/Commercialisation of Agricultural Technology (Upscaling of existing component IPR HQ)
8. Indo US Knowledge Initiative
9. National Agricultural Innovative Project, New Delhi
10. National Fund for Basic and Strategic Research, New Delhi

Terminology

AB line: The imaginary reference line set for each field that a tractor/sprayer guidance system

Absorption Losses: Loss of water from a canal or a reservoir by capillary action and percolation and in case of canal during the process of delivery.

Accuracy (of GPS receivers): The measure of closeness of an object's actual (true) position to the position obtained with a GPS receiver. Accuracy levels are used to rate the quality of GPS receivers.

Acid soil: A soil which is deficient in available bases, particularly Ca and which give an acid reaction when tested by a standard method.

Acid Soil: A soil with an acid reaction, a pH less than 7.0.

Acre :(43560 sq. ft) an area of land about 220 feet long and 198 feet wide.

Acre foot water: The amount of water that would cover an area of land to a depth of one foot assuming no seepage evaporation and run off.

Acre inch day: Term used principally in irrigated section of united state for measuring quantity of flow of water. It is equal to a flow which will cover one acre to a depth of one inch in a 24 hours period or 0.042 cubic feet per second.

Acre inch: It is a measure of quantity of flow of water and is equal to the flow which will cover one acre to a depth of one inch.

Acre: A parcel of land, containing 4,840 square yards or 43,560 square feet

Active Down Force (sometimes displayed as "down force margin") - A system which automatically adjusts the force in the air spring circuit based on soil condition information gathered from the row unit gauge wheel sensors.

Active sensing systems: Sensing systems that generate a signal, bounce it off an object, and measure the reflected signal.

Actuator: A device used in variable-rate application that responds to controller signals to regulate the amount of material applied to a field.

Adaptive Sampling: Dynamic sampling plan that changes over time based on actual field conditions and analysis results; often affects the number and location of samples.

Adiabatic: A condition in which heat is neither gained nor dissipated.

Adobe soil: These soils are formed by the broken material of rocks transported by both wind and water.

Aerial Imaging: Photos taken, or images collected, from aircraft to assist growers and consultants in determining variations within an area of interest such as a farm field.

Aerial photography: Photos taken from airplanes to assist growers to determine variations within an area of interest such as a field.

Aerobic (i) Having molecular oxygen as a part of the environment. (ii) Growing only in the presence of molecular oxygen, such as aerobic organisms. (iii) Occurring only in the presence of molecular oxygen (said of chemical or biochemical processes such as aerobic decomposition).

Agar: A substance made from seed weed and used in the solid culture.

Agriculture anomaly: An agronomic (vegetation or soil) deviation or inconsistency in excess of "normal" variation from what one would expect to observe.

Agriculture: It is an art, science and business of raising crops and rearing of animals through exploring the natural resources with the coordination of socio economic infrastructure to meet the basic necessities of life.*i.e.*food, feed, fiber and shelter.

Air (-filled) porosity: The fraction of the bulk volume of soil that is filled with air at any given time or under a given condition, such as a specified soil-water content or soil-water matric potential.

Albedo: The ratio of the amount of solar radiation reflected by a body to the amount incident upon it, often expressed as a percentage, as, the albedo of the earth is 34 per cent.

Allelopathy: Phenomenon involving the release of certain chemicals from plant parts into the environment which may when present in sufficient amounts, inhibit or suppress the germination or growth of the plants in the neighborhood.
Alluvial soil: These are the soils which are formed by the deposition of broken material of rocks transported and deposited by water of streams and rivers.

Altitude: Height from sea level

Anion exchange capacity: The sum of exchangeable anions that a soil can adsorb. Usually expressed as centimoles, or millimoles, of charge per kilogram of soil (or of other adsorbing material such as clay).

Anti-Spoofing: Process of encrypting the L2 signal to prevent unauthorized transmissions of false GPS signals.

Apiary: Colonies of bees in hives and other beekeeping equipment for the production of honey. Application Rate: Amount of seed distributed, expressed as a number, mass or volume of seed per unity of length or surface.

Application losses: Water losses through percolation or run off.

Arable farming: The term arable farming refers to system in which only crops that require cultivation of the soil are grown.

Arboriculture: Intensive cultivation of individual trees possibly for fruits gums and resins.

Arid region: The region where total rain fall is less than natural evapo – transpiration rate.

Aridity: It is the characteristic of a region where there is low average rain fall or 100 per cent available water. It is permanent feature of region.

Auger: Spiral device on a shaft used to move grain through a tube.

Available water: The water retained in a soil which represents the difference between field capacity and the permanent wilting percentage is called available water.

Barani soil :When the source of irrigation to crop is only the rain water that is known as Barani soil.

Bare fallow: Complete inversion and incorporation of residues for maximum decomposition, done to prevent the growth of all vegetation; usually associated with summer fallow.

Base period: Period of time in days from the first watering of crop before sowing and the last watering before harvesting.

Basic seed : Is the progeny of pre – basic seed produced so as to maintain genetic purity and identity.

Basin: Flat area of land surrounded by low ridges or bunds

Biological yield :It is the total dry matter produced by a plant as a result of photosynthesis and nutrient uptake minus that lost by respiration.

Biomass: The weight of living organisms (plants and animals) in an ecosystem, at a given point in time, expressed as fresh or dry weight.

Blind hoeing : Hoeing before a crop germinates.

Bolting: Formation of elongated stem or seed stalk, it is usually takes place during the second season of the growth in biennial plants.

Botanical variety :When a group of plant occurring in nature is different from the general species originally described and the botanical binomial name is not enough to identify it is called as botanical variety.

Broadcasting: Manual spreading of seed in the field and mixing of the spread seed by ploughing or planking the field.

Broadcasting: Random scattering of seeds over the surface of the ground. If the seed is to be covered, this is done as a separate operation, usually with a spike-tooth harrow.

Bt crops: Crops that are genetically engineered to carry a gene from the soil bacterium Bacillus thuringiensis (Bt). The bacterium produces proteins that are toxic to some pests but non-toxic to humans and other mammals. Crops containing the Bt gene are able to produce this toxin, thereby providing protection for the plant. Bt corn and Bt cotton are examples of commercially available Bt crops.

C3 plants: Plants which fix CO_2 in three C molecule and do not use temperature and water as efficiently as C4 plant. e.g wheat, rice, cotton.

C4 plant : Plants which fix CO_2 in to a four C molecules.e.g sugarcane, maize, sorghum.

Capillary Water: It is that water which is held by surface forces (adhesion, cohesion, surface tension) or films around the particles in angles between them and in capillary pores.

Capillary Water: It is the soil water in excess of hygroscopic water. This exists in the pore space of the soil by surface tension or molecular attraction against gravitational forces. It is only water available for plant growth and development.

Cash Crop: Any crop that is sold off the farm to yield ready cash.

Catchment's Area: The area which drains the rain water falling on it, via streams and rivers, eventually to the sea or into a lake.

Cation Exchange Capacity (CEC): Represents the total quantity of negative charge that is available in the soil to attract positively-charged ions in the soil solution.

Cereal Crops: A cereal is defined as crop grown for its edible seed. These crops are also known as grain crops *e.g.* wheat, Rice, Maize *etc.*

Certified Seed: Seed grown from pure stock which meets the standards of certifying agency (usually a state government agency). Certification is based on germination, freedom from weeds and disease, and trueness to variety.

Certified Seed: It is the progeny of basic seed and is produced by registered growers of seed producing agencies.

Clayey Soil: A soil is known as clayey which contain at least 30 per cent clay particles and in most cases not less than 40 per cent,usually it contain 45 per cent clay, 30 per cent silt and 25 per cent sand.

Climate : Aggregate of atmospheric condition over a long period of time.

Clone : A cultivar propagated by vegetative method is called a clone.

Colluvial Soil : Are those which are form from the material transported by the force of gravity.

Command Area : Area which can be economically irrigated by an irrigation system.

Commercial Farming System : In this type of farming system, crops are raised on a commercial scale for marketing.

Companion Crops : The two crops grown together are called companion crops. *e.g.* Berseem and barley.

Complete Fertilizer: A fertilizer containing the three macro nutrients (Nitrogen, Phosphorous, and Potassium) in sufficient amounts to sustain plant growth.

Compost: A manure derived from decomposed plant remains usually made by fermentation, waste plant material under controlled conditions. Compost usually used in green houses to enrich the soil either dungs as surface.

Compost: Organic residues, or a mixture of organic residues and soil which have been piled, moistened, and allowed to undergo biological decomposition. Mineral fertilizers are sometimes added.

Condiment Crops : Crops which are grown and consumed as condiments *e.g.* coriander, mint.

Conidia :One celled asexual spores in certain fungi.

Consumptive Use of Water :Evapo-transpiration plus the water assimilated by various plant metabolic processes. As the water consumed in plant metabolism is very small, consumptive use and evapo-transpiration are considered almost equal.

Contour Line: A line used to represent the same value of an attribute (elevation or yield). Contouring – interpolation method used to distinguish between different levels of an attribute (elevation, fertility, yield).

Contour Map: Yield map that combines dots of the same intensity/yield level by interpolating (or kiging).

Cooperative: An organization formed for the purpose of production and marketing of goods or products owned collectively by members who share in the benefits. Most common examples in agriculture are canneries and creameries.

Cover crop: Close-growing crop, that provides soil protection, seeding protection, and soil improvement between periods of normal crop production, or between trees in orchards and vines in vineyards. When plowed under and incorporated into the soil, cover crops may be referred to as green manure crops.

Cover Crops : The crops, which are planted to cover the ground and to reduce the soil erosion and nutrients losses by leaching. *e.g.* grasses and rye.

Crassulation Acid Metabolism (CAM) Plants :CAM plants fix CO_2 in four carbonic acid as do the C4 plants *e.g.* pine apple.

Critical Period of Competition :During the crop period there is a certain time when crop plants are most sensitive to competition by weeds, this time is known as the critical period of competition.

Critical Threshold Level (CTL) : A weed, insect pest density capable of causing significant damage to crop is termed as critical threshold level.

Crop : A crop is a community of plants grown under field condition for its economic value.

Crop residue: Portion of plants remaining after seed harvest; refers mainly to grain crop residue, such as corn stover, or of small-grain straw and stubble.

Crop Rotation : Is the strategy of raising crops from a piece of land in such an order or succession that the fertility of land suffers minimally and the farmer's profits are not reduced.

Crop Rotation: More or less regular recurrent succession of different crops on the same land for the purpose of maintaining good yields.

Crop Sensors: Optical crop sensors used to measure and/or quantify crop health or evaluate crop conditions by shining light of specific wavelengths at crop leaves, and measuring the type and intensity of the light wavelengths reflected back to the sensors.

Crop Sensors: Optical crop sensors used to measure and/or quantify crop health or evaluate crop conditions by shining light of specific wavelengths at crop leaves, and measuring the type and intensity of the light wavelengths reflected back to the sensors.

Crop Variety: The distinctive name of the crop type or the named, specific characters used to identify the crop.

Crop Water Requirement : The amount of water required to raise a crop to maturity with in a given period of time.

Crop Year (commonly referred to as Growing Season): The period within which a crop is normally grown, regardless of whether or not it is actually grown, and designated by the calendar year in which the crop is normally, harvested.

Cropping Intensity : The term cropping intensity refers to the ratio of actual cultivated area to total farm area over a year.

Cropping intensity: Refers to the no. of crops which are raised during the year.

Cropping Pattern: It is a general cropping system followed or practiced by the farmers in an ecological zone.

Cropping Scheme : Allocation of an area to different crops being grown on a particular farm in a year.

Cultivator :Which only cut and stirr the soil.

Define Necrosis : Death of organs of a plant, either as blight or death of tissue in localized areas, usually inside fruit and stems or die back or death of stems or branches.

Define Olericulture : Branch of horticulture which deals with cultivation of vegetables.

Delta of Water : The depth of irrigation water required for the full crop period.

DEM -(Digital Elevation Model): a digital representation of the elevation of locations on the land surface. A DEM is often used in reference to a set of elevation values representing the elevations at points in a rectangular- grid on the Earth's surface. Some definitions expand DEM to include any digital representation of the land surface, including digital contours.

Denitrification: Reduction of nitrogen oxides (usually nitrate and nitrite) to molecular nitrogen or nitrogen oxides with a lower oxidation state of nitrogen by bacterial activity (denitrification) or by chemical reactions involving nitrite (chemodenitrification). Nitrogen oxides are used by bacteria as terminal electron acceptors in place of oxygen in anaerobic or microaerophilic respiratory metabolism.

Determinate Plants :Those plant which initiate their reproductive stage after completing vegetative growth, *e.g.* wheat, barley.

Dicots : Dicots have two cotyledons and reticulate leaf venation.

Differential Global Positioning System (DGPS): a method of using GPS which attains the position accuracy needed for precision farming through differential correction.

Diversified Farming: This is an expanded type of farming system in which varieties of crops are produced and many types of animals are reared.

Dobari Crops: A crop grown on residual moisture after the harvest of rice.

Dormancy: Seed dormancy is the state of inhibited germination of seeds with viable embryos in condition conducive to plant growth.

Dormant seeding: The practice of planting seed during the late fall or early winter after temperatures become too low for seed germination to occur until the following spring.

Double Crop: Two different crops grown on the same area in one growing season.

Drainage: It is the removal of excess surface or ground water from the root zone of a crop by means of surface or sub – surface drains.

Drainage: The removal of excess surface water or excess water from within the soil by means of surface or sub-surface drains.

Drilling: The process of opening the soil to receive the seed, planting the seed and covering it in a single operation.

Dry farming: In which crops and livestock are raised on land which does not receive sufficient rainfall for water intensive crops and no irrigation facilities are available fall into this category.

Dry Land Farming: The practice of crop production without irrigation.

Duty of water: The relationship between irrigation water flow and its commanded area where crops mature fully with that amount of water within a base period is called duty of water.

Earthing up: The operation of pulling up soil from the center of crop rows to the bottom of the plants, this helps in uprooting weeds and supporting to plants.

Economic yield: The economically important part for which a particular crop is grown.

Effective rainfall: It is the part of the rainfall which forms a portion of the water requirement of a crop or which can be used by crop.

Eolian soil: The soil which is formed by the material transported by winds from one place to another is called eolian soil.

Epigeal germination: It is derived from two words epi "above" and geas "earth". In this type of germination the cotyledons come out above the soil surface and generally turn green and act as first foliage leaves. *e.g.* bean,cotton.

Erosion: The wearing away of the land surface, usually by running water or wind.

Evapotranspiration:It is the total loss of water due to its evaporation from land, plant and water surfaces and transpiration by vegetation per unit area per unit time.

Exhaustive crops: Crops, which feed heavily on the soil and deplete soil nutrients *e.g.* sorghum, tobacco.

Extensive farming: In this type of farming large areas are used with minimum expenditure or attention to efficient use of other resources.

Fertilizer Any organic or inorganic material of natural or synthetic origin (other than liming materials) that is added to a soil to supply one or more plant nutrients essential to the growth of plants.

Fiber crops: The crops, which are grown for their fiber and are used in making textiles, ropes. *e.g.* jute, sun hemp, cotton.

Field capacity: The amount of water retained by soil after drainage of saturated soil by gravitational force is called field capacity.

Field Capacity: The moisture content of soil in the field as measured two or three days after a thorough wetting of a well-drained soil by rain or irrigation water.

Forage crops: Those crops, which are grazed by animals and harvested for green chop, hay, silage are classified as forage crops *e.g.* maize and sorghum.

Forage: Vegetable matter, fresh or preserved, which is gathered and fed to animals as roughage (*e.g.*, alfalfa hay, corn silage, or other hay crops).

Fruit farming: In which orchards are planted and the objectives are to maximize fruit production, enhance quality and increase income.

Functional Allelopathy: It is the case when the chemicals are toxic after being transformed by microorganism.

Garden crops: Vegetable crops, which are grown for their edible leaves, shoots, flowers, fruit and seed. *e.g.* cabbage and okra.

Geographic Data: Data that contains information about the spatial location (position) and the attribute being monitored (yield, seed population, *etc.*). Also referred to as spatial data.

Georeferencing: The process of adding geographic data to yield data or other field attributes either in real-time (on-the-go) or by post-processing or the process of associating data points with specific locations on the earth's surface.

Geo-Stationary Satellite: Space vehicle in an orbit that keeps it over the same location on the earth at all times.

Germination (1) Resumption of active growth by the seed embryo, culminating in the development of a young plant. (2) In seed laboratory practice: emergence and development from the seed embryo of those essential structures, which, for the kind of seed in question, are indicative of the ability to produce a normal plant under favorable conditions.

Germination: Is the emergence and development from the seed embryo of those essential structures which, for the kind of seed provided, indicate the ability to produce a normal plant under favourable conditions.

Gibberellins: Plant growth stimulating chemicals which are able to induce a number of effects on plants.

Grain moisture content: Moisture content (MC) is the weight of water contained grain. The moisture content is generally reported on the wet basis meaning the total weight of the grain including the water.

Grassland farming: These systems are mainly concerned with growing grasses for consumption by livestock kept for milk or meat production.

Gravitational potential: It is produced by gravitational forces operating on soil water.

Gravitational water: Is the water in excess of hygroscopic and capillary water that percolates through the soil under the action of gravity if favourable conditions for water drainage are provided.

Green manure crops: Some crops are grown and ploughed in the soil in green form in order to improve soil fertility *e.g.* Berseem, Guara,Dhaincha *etc.*

Green Manure: Any crop or plant grown and plowed under to improve the soil, by addition of organic matter and the subsequent release of plant nutrients, especially nitrogen.

H.I. = economic yield x 100/Biological yield

Hard pan: A hard semi impervious layer usually developed due to continuous ploughing to a depth of about 15 cm, with cultivators, or with continuous deposition of salts due to soil or surface irrigation water.

Harvest index: The ratio of grain weight to total plant weight in a cereal crop.

Herbaceous: Plants with soft and easily vulnerable body parts.

Herbicide-tolerant crops: Crops that have been developed to survive application(s) of particular herbicides by the incorporation of certain gene(s) either through genetic engineering or traditional breeding methods. The genes allow the herbicides to be applied to the crop to provide effective weed control without damaging the crop itself.

Herbs: Are plants of small to medium height and canopy.

Humus: The well decomposed, relatively stable portion of the organic matter in a soil.

Hybrid vigour: Qualities in a hybrid not present in either parent.

Hydrophyte: Plant which grows in water, or which loves water.

Hygroscopic water: Water attached to soil particles by loose chemical bonds and does not move by the action of gravity or capillary force.

Ideotype: An ideal plant type developed through breeding.

Indeterminate plants: In these plants, the vegetative and reproductive stages continue simultaneously *e.g.* okra, tomato. Soybean is the only crop, which has determinate and indeterminate as well as semi – determinate growing types.

Inoculant: The bacteria containing material used to introduce N – fixing Rhizobium bacteria into soil.

Insecticide resistance: The development or selection of heritable traits (genes) in an insect population that allow individuals expressing the trait to survive in the presence of levels of an insecticide (biological or chemical control agent) that would otherwise debilitate or kill this species of insect. The presence of such resistant insects makes the insecticide less useful for managing pest populations.

Integrated weed management: The concept of IWM involves the planned use of all possible direct and indirect measures rather than relying on a single method to combat weeds.

Inter cropping: Growing of two or more crops together on the same field, where one crop (main crop) planted in rows first and then another crop (intercrop) is planted in between the rows. These crops remain in association for a shorter time. These crops may or may not be planted and harvested at the same time e.g in Sugar cane planting of onion, garlic *etc.*

Interception: When the drops of rainfall or precipitation are intercepted by plant leaves it is called interception.

Irrigated soil: The soil, which receives irrigation water from well and tube wells, are known as irrigated or chahi soil.

Irrigation efficiency: It is a term used to indicate how efficiently the available water supply is being used for crop production.

Irrigation scheduling: It refers to the number of irrigations for a crop and their timing.

Irrigation water requirement: The quantity of water required for successful crop production exclusive of precipitation, ground water and other natural resources.

Irrigation: Irrigation is the artificial application of water to soil or crop plants to assist crop production.

Kera: Manual sowing of seeds in lines in furrows.

Kharif crops: Those crops, which are planted in the summer month from the March to July and harvested in autumn and winter, are called Kharif crops *e.g.* rice and cotton.

Lacustrine soil: When the material transported by streams and river, if deposited in lakes the soils are called Lacustrine soil.

Land Classification: (land capability class) The classification of units of land for the purpose of grouping soil of similar characteristics, in some cases showing their relative suitability for some specific use.

Landscape: A collective term for all the natural features (such as fields, hills, forests, water, *etc.*) that distinguish one part of the earth's surface from another part. Usually used in reference to that land or territory which the eye can comprehend in a single view, including all its natural characteristics.

Latitude: The angular distance north or south of earth equator.

Line: A cultivar propagated by seed is called a line.

Livestock and poultry farming: This category includes farming system in which various kinds of livestock are reared for meat, milk, wool and eggs.

Loamy soil: The soil is more or less than midway clay and sandy soil and ideal loam soil may defined as a mixture of sand, silt and clay particles which shows the properties of sandy, silt and clayey in equal proportion.

Lodging: The bending or breaking over of a plant before harvesting.

Long day plants: Plants which change vegetative to reproductive stage by producing flowers and fruits, when the days become longer. *e.g.* Carrot, Radish *etc.*

Matric potential: It is produced by capillary and surface forces.

Maximum potential soil moisture deficit: Is the greatest value of potential soil moisture deficit attained during the growth of a crop.

Monocots: These plants have one cotyledon and parallel leaf venation.

Muck soil: If the quantity of organic matter exceeds 20 per cent but less than 50 per cent are called muck soil.

Mulch: Any material or practice which is used to check the loss of water by evaporation is called mulch.

Multiple cropping: It is growing of two or more crops in a year from the same piece of land

Narcotic or drug crops: This category includes those crops, which have some narcotic and drug value *e.g.* poppy, tobacco.

Natural erosion: The erosion of the soil under natural condition.

Natural soil: Strictly speaking a soil having pH of 7, in practice a soil having PH 6.6 and 7.3.

Net plot: Area from which yield and other characteristics are measure. It is also known as the net area of the plot.

Nitrate toxicity: A variety of conditions in animals, resulting from ingestion of feed high in nitrate; the toxicity actually results when nitrate (NO_3) is reduced to nitrite (NO_2) in the rumen.

Nitrification: Biological oxidation of ammonium to nitrite and nitrate, or a biologically induced increase in the oxidation state of nitrogen.

Nitrogen (N): An inert gas that makes up about four-fifths of the air. Nitrogen for commercial purposes can be "fixed" synthetically from the atmosphere by several processes. A nutrient critical to plant growth.

Nitrogen Cycle: The sequence of transformations undergone by nitrogen in its movement from the free atmosphere into and through soils, into the plants, and eventually back. These biochemical reactions are largely involved in the growth and metabolism of plants and microorganisms.

No tillage crop(zero tillage crop): Crop grown with out any tillage to prepare seed bed or row.

Node A slightly enlarge portion of stem where leaves and bud arises and where branches originate.

Nucleus seed: Seeds obtained from selected individual plants of a particular variety which needs to be purified and multiplied in such a way as to maintain its genetic purity.

Nut cycle: The regeneration/cycling of nutrients.

Nutrients budget: A quantitative data of the major nut flowing to retained within the discharge from the system.

Nutrients: The food for microbial and plant life mainly composed of nitrogen and phosphorous but also of potassium, Mg, Fe, Ca,Co,Cu, Zn and others elements.

Oilseed crops: These are the crops, which are grown for the purpose of extracting oil from their seed *e.g.* mustard and groundnut.

Organic agriculture: A concept and practice of agricultural production that focuses on production without the use of synthetic inputs and does not allow the use of transgenic organisms. USDA's National Organic Program has established a set of national standards for certified organic production which are available online.

Organic Fertilizer: Any fertilizer material containing plant nutrients in combination with carbon.

Osmotic potential: It is also called solute potential. It is produced by various solutes in soil water.

Particle density: The density of the soil particles, the dry mass of the particles being divided by the solid (not bulk) volume of the particles, in contrast with bulk density. Units are Mg m-3.

Particle size: The effective diameter of a particle measured by sedimentation, sieving, or micrometric methods.

Peat soil: If the quantity of organic matter is more than 50 per cent is called peat soil.

Percolation: Downward movement of water with in the soil profile.

Permanent wilting percentage: The soil water content at which plants can no longer extract sufficient water from the soil for their growth is called permanent wilting percentage.

Plant development: Plant development is the progress of plant from germination to maturity through a series of stages.

Plant growth: It is the increase in the dry weight of a plant over time mainly as a consequence of photosynthesis.

Plough pan: A dense, compacted layer about 5 to 7 cm thick formed beneath the surface soil by repeated ploughing in the same path.

Plough: Its function is to cut, stir, invert, and pulverize the soil.

Pore space: It is a space between soil particles occupied by air and water; it is largely controlled by the texture of soil.

Potential Evapotranspiration: Is defined as the amount of evaporation occurring from an extensive area of a short, green growing crop completely covering the ground and well supplied with water.

Potential soil moisture deficit: It is the difference between a crop potential evapotranspiration and the amount of rainfall received by a crop plus the quantity of water delivered to it in irrigation.

Pre – basic seed: It is the progeny of nucleus seed, and is handled so as to maintain specific genetic purity and identity as completely as possible.

Precision Farming - managing crop production inputs (seed, fertilizer, lime, pesticides, *etc.*) on a site-specific basis to increase profits, reduce waste and maintain environmental quality.

Pressure potential: It is produced by actual hydrostatic pressure.

Puddling: Ploughing in standing water to create a shallow hard pan at a 10 to 15 cm depth, which helps to increase water-holding capacity and reduce moisture losses by percolation.

Pulses or grain legumes: The crops belonging to Leguminoseae family are grown for their edible seed e.g chick pea, lobia.

Rabi crops: These crops are planted in winter from October to December and harvested in summer from March to May *e.g.* wheat,

Readily available water: The portion of the available water that is most easily extracted by a plant is called readily available water.

Relay crops: A relay crops is one which is planted as a second crop after the first crop has reached its reproductive stage of growth but before it is ready for harvest. *e.g.* planting of sugar cane in sugar beet.

Relief: When rising ground or mountains running at right angle to the prevailing wind.

Root and tuber crops: These are vegetable crops grown for their under ground parts like roots, bulbs, rhizomes, corms and stem tubers *e.g.* carrot and onion.

Rumber: The process in which planker or leveler is used to conserve moisture at watter condition before the preparation of land after rauni.

Run off: When water flows out the field by breaking the bunds of the field or flows to the sloppy areas from the high level is called run off.

Sandy soil: the soil which contains 2.00 to 0.2 mm diameter soil particles, it contain 85 per cent sand, and >15 per cent, silt and clay. These soils are poor in plant material.

Saturation capacity:This term refers to the amount of water present in the soil when it is completely saturated with water.

Scarification: Any physical or chemical treatment that makes the seed coat permeable is known as scarification.

Secondary Nutrients : The secondary plant foods include calcium, magnesium and sulfur. Lesscritical elements required in smaller amounts for plant growth than nitrogen, potassium and phosphorous.

Secondary tillage: Tillage that works the soil to a shallower depth than primary tillage, providing additional pulverization; levels and firms the soil, closes air pockets, kills weeds.

Seed certification: Is the process to secure, maintain and make available high quality seed and propagating materials of superior crop varieties, so grown and distributed as to ensure desirable standards of genetic identity, physical purity and quality attributes.

Seed certification: Refers to the system of maintaining the genetic purity and quality of seed.

Seed dressing: The chemical treatment of seeds particularly cereals, with fungicides and some time insecticides to protect them against soil and seed borne disease and pest.

Seed Hybrid: - The identification given to seed by a seed company; a seed that has been developed by selective genetics and cross-breeding.

Seed vigor: Those seed properties that determine the potential for rapid uniform emergence and development of normal seedlings under both favorable and stress conditions.

Seedling A young plant that has grown from a seed

Seeling: Mechanical weeding process carried out by ploughing with bullocks or tractors.

Seepage: Losses through the beds and banks of canals and water channels take place mainly by unlined canals.

Sett: A piece of seed cane with two – four buds(eyes).

Sheet Erosion: The gradual, uniform removal by water of the earth's surface, without the formation of hills or gullies.

Short day plants: Plants which changes from vegetative to reproductive stage and produces flowers and fruits, when the days become shorter.

Shrubs: Are bushy plants with medium to tall height and canopy.

Silage crops: Those crops, which are harvested when still green and succulent and are fed directly to animals with out curing, are called silage crops. *e.g.* Berseem, Shaftal, maize.

Soil test : A chemical, physical, or biological procedure that estimates the suitability of a soil to support plant growth.

Soil Texture: Refers to the coarseness or fineness of a soil. It is determined by the relative proportion of various sized particles (sand, silt, and clay) in a soil.

Soil: A natural body developed from variable mixtures of broken and weathered minerals and decaying organic matter which covers the earth crust in thin layers and supplies proper amount of nutrients and air water and mechanical support to plants.

Stratification: Is the practice of exposing imbibed seeds to cool temperature conditions for a few days prior to germination in order to break their dormancy.

Subsistence farming: In which basic necessities like food, clothing, and shelter are produced for the family to live on.

Sugar crops: These are the crops, which are grown for sugar purposes. *e.g.* sugar cane and sugar beet.

Summer Fallow: Land plowed up (usually in spring) and left unseeded through the summer. this is done to let the land air out and rest until fall, when it is worked up and planted to a crop of grain. May also be done to beak down organic matter or kill weeds.

Temporary wilting percentage: Soil water content at which plants wilt during the hot windy part of the day but regain turgidity during the cooler part of the day is called TWP.

That part of the crop for which crop is grown *e.g.* stem in sugarcane, root in the case of sugar beet.

Tillage: The mechanical manipulation of the soil profile for any purpose; but in agriculture it is usually restricted to modifying soil conditions and/or managing crop residues and/or weeds and/or incorporating chemicals for crop production.

Tiller: To put forth shoots other than the mother shoot from roots axis.

Tilth: The physical condition of soil is called Tilth.

Top Dressing: Lime, fertilizer, or manure applied after the seedbed is ready, or after the plants are up.

Topping: Topping in tobacco is the removal of the terminal bud with or without some of the small top leaves just before or after the appearance of the flower head.

Topsoil: The layer of soil used for cultivation, which usually contains more organic matter than underlying materials.

Transpiration ratio: It is the ratio of the weight or volume of water transpired by the plant during its growth period to the weight of dry matter produced by the plant.

Transpiration: It is the process of loss of water from living plants.

Trench layering: An asexual reproductive method of plant propagation involving laying down the whole stem, the new shoots are thus forced to push their way through a layer of soil which prevents the bark from coloring and favors root formation.

Truch gardening: Growing of crops like potato onion and cabbage on large scale for distinct market.

Truck farming: It refers to the system in which the bulk of the out puts are produced and marketed.

True Allelopathy: When the allelochemical is toxic in their original form is called true Allelopathy.

Variety: In general the term variety has been used to refer to a group of similar plant with in a particular species that is distinguished by one or more then one character and given the name.

Vegetation Indexes: A tool for identifying the levels in health of plant biomass. A vegetation index can be used to assess or predict plant characteristics such as leaf area, total plant material, and plant stress. A vegetation index reduces several wavelengths of sensor data into a single number.

Vegetative cover: A soil cover of plants irrespective of species.

Vernalization and chilling: Many biennials and temperate annuals, as well as certain fruit trees, require exposure to cold temperature before they can flowers. This is known as Vernalization requirement for annuals and biennials and chilling requirement for fruit trees.

Vines: Are plants, which have tender stems, and requires some support for upward growth.

Water holding capacity: Ability of soil to retain water.

Water potential: Refers to the chemical potential of water.

Water Table: The upper limit of the part of the soil or underlying rock material that is wholly saturated with water. In some places an upper or perched water table may be separated from a lower one by a dry zone.

Watter: Describes the condition when soil moisture level is suitable for cultivation.

Weather: It is a state of atmosphere at any time; it is combined effect of many things such as heat, cold.

Weathering: The process by which soil disintegrates and decomposes, eventually producing soil particles by exposure to the physical and chemical effects of atmospheric agents.

Weed: any plant growing out of its proper place.

Wilting point: Water content of a soil when indicator plants growing in that soilwilt and fail to recover when placed in a humid chamber.

Wilting point: The point at which the water content of a soil reaches such a level that it is firmly held by soil and unavailable to plant roots, so that the plants wilt permanently and die.

Yield Monitor - A yield-measuring device installed on harvest machines. Yield monitors measure grain flow, grain moisture, and other parameters for real-time information relating to field productivity.

Zaid Kharif crops: These are the crops, which are planted in August to September and harvested in December to January *e.g.* toria.

Zaid rabi crops: These are crops, which are planted in February and harvested in May-June *e.g.* tobacco.

Zone Management – The information-based division of large areas into smaller areas for site specific management applications.

References

Ashawa G.H. Irrigation Engineering (1999), New Age International Publishers, New Delhi.

Bais H P, Vepachedu R, Gilroy S, Callaway RM and Vivanco JM (2003b). Allelopathy and exotic plant invasion: from molecules and genes to species interactions. *Science* 301,1377-1380.

Bais HP, Walker TS, Kennan AJ, Stermitz FR and Vivanco JM (2003a). Structure-dependant phytotoxicity of catechins and other flavonoids: flavonoid conversions by cell-free protein extracts of *Centaurea maculosa* roots. *Journal of Agriculture and Food Chemistry* 51, 897-901.

Bais HP, Walker TS, Stermitz FR, Hufbauer RA and. Vivanco JM (2002). Entaniomeric-dependent phytotoxic and antimicrobial activity of (+-) catechin. A rhizosecreted racemic mixture from spotted knapweed. *Plant Physiology* 128, 1173-1179.

Balasubramaniyan P. and S.P. Palaniappan Principles and Practices of Agronomy (2004), Agrobios (India), Jodhpur.

Barnes J P and Putnam A R (1987). Role of benzoxazinones in allelopathy by rye (*Secale cereale* L.). *Journal of Chemical Ecology* 13, 889-905.

Barnes JP and Putnam A R (1983). Rye residues contribute weed suppression in no-tillage cropping systems. *Journal of Chemical Ecology* 9, 1045-1057.

Barney J N and DiTommasso A (2003). The biology of Canadian weeds. *Artemisia vulgaris* L. *Canadian Journal of Plant Science* 83, 205-215.

Barney J N, Hay A and Weston L A (2005). Isolation and characterization of allelopathic volatiles from mugwort (*Artemisia vulgaris* L.) foliage. *Journal of Chemical Ecology* 31, 247-265.

Beerling D J, Bailey JP and Conolly AP (1994). *Fallopia japonica* (Houtt.) Ronse Decraene. Biological Flora of the British Isles. *Journal of Ecology* 82, 959-979.

Bell D T and Muller C H (1973). Dominance of California annual grasslands by *Brassica nigra. American Midlands Naturalist* 90, 277-299.

Bertholdsson N (2004). Variation in allelopathic activity over 100 years of barley selection and breeding. *Weed Research* 44, 78-86.

Bertin C, Paul R N, Duke, SO and Weston, L A (2003). Laboratory assessment of the allelopathic potential of fine leaf fescues (*Festuca rubra* L.). *Journal of Chemical Ecology*8:1919-1937.

Blum U (1998). Effects of microbial utilization of phenolic acids and their phenolic acid breakdown products on allelopathic interactions. Journal of Chemical Ecology 24: 685-708.

Blum U (2002). Soil solution concentrations of phenolic acids as influenced by evapotranspiration. Abstracts of the Third World Congress on Allelopathy. p.56.

Blum U and Shafer, S R (1988). Microbial populations and phenolic acids in soil. *Soil Biology and Biochemistry* 20:793-800.

Blum U. (1995). The value of model plant-microbe-soil system for understanding processes associated with allelopathic interactions. *In* Allelopathy, Organisms, Processes and Applications. eds. Inderjit, Dakshini and Einhellig. ACS Symposium Series 582.Washington DC. pp. 127-131.

Burgos N R and Talbert R E (2000). Differential activity of allelochemicals from *Secale cereale* in seedling bioassays. *Weed Science* 48, 302-310.

Burgos N R, Talbert R E and Mattice J D (1999). Cultivar and age differences in the production of allelochemicals by *Secale cereale. Weed Science* 47, 481-485.

Callaway R M and Aschehoug E T (2000). Invasive plants versus their new and old neighbors: A mechanism for exotic invasion. *Science* 290, 521-523.

Chevallier A (1996). The Encyclopedia of Medicinal Plants : A Practical Reference Guide to More Than 500 Key Medicinal Plants and Their Uses. 336 pp., DK Publishing, New York.

Chou C H and Lin H J (1976). Autointoxication mechanism of *Oryza sativa*. I. Phytotoxic effects of decomposing rice residues in soil. *Journal of Chemical Ecology* 2, 353-367.

Czarnota M A (2001). Sorghum (*Sorghum* spp.) root exudates : production, localization, chemical composition, and mode of action. Ph.D. thesis, Cornell University.

D.C. Gopal Chandra, Fundamentals of Agronomy (1990), Oxford and IBH Publishing Co. Pvt. Ltd., New Delhi.

Das P.C. Manures and Fertilizers (1999), Kalyani Publishers, Ludhiana.

deCandolle M A P (1832). *Physiologie Vegetale. Bechet Jeune Library Faculty Medicine Paris* 3, 1474-1475.

Dewey SL (1986) Effects of the herbicide atrazine on aquatic insect community structure and emergence. Ecology 67(1):148-162.

Dilday R H, Nastasi P and Smith R J J (1991). Allelopathic activity in rice (*Oryza sativa* L.) against ducksalad (*Heteranthera limosa*). *In* Sustainable Agriculture for the Great Plains, Symposium Proc. M J S J D Hanson, D A Ball, C V Cole. USDA, Washington, DC. pp 193-201.

Duke S O (1986). Naturally occurring chemical compounds as herbicides. *Reviews in Weed Science* 2, 15-44.

Duke S O, Dayan F E and Romagni J (2000). Natural products as sources for new mechanisms of herbicidal action. *Crop Protection,* 19, 572-575.

Duke S O, Dayan F E, Rimando A M, Shrader K, Aliotta G, Oliva A and Romagni J G (2002). Chemicals from nature for weed management. *Weed Science* 50, 138-151.

Duke S O, Scheffler B E, Dayan F E, Weston L A and Ota E (2001). Strategies for using transgenes to produce allelopathic crops. *Weed Technology* 15, 826-834.

Duke SO (1990) Overview of herbicide mechanisms of action. Environmental Health Perspectives 87:263-271.

Einhellig F A and Rasmussen J A (1989). Prior cropping with grain-sorghum inhibits weeds.*Journal of Chemical Ecology* 15, 951-960.

Feurtado JA, Ambrose SJ, Cutler AJ, Ross AR, Abrams SR, Kermode AR (February 2004). "Dormancy termination of western white pine (Pinus monticola Dougl. Ex D. Don) seeds is associated with changes in abscisic acid metabolism". Planta 218 (4): 630– 9. doi:10.1007/s00425-003-1139-8. PMID 14663585

Folmar LC, Sanders HO, Julin AM (1979) Toxicity of the herbicide glyphosate and several of its formulations to fish and aquatic invertebrates. Archives of Environmental Contamination and Toxicology 8:269-278.

Fortuna A M, de Riscala E C, Catalan C A N, Gedris T E and Herz W (2002). Sesquiterpene lactones and other constituents of *Centaurea diffusa. Biochemistry and Systems Ecology* 30, 805-808.

Gonzalez V M, Kazimir J, Nimbal C, Weston. A and Cheniae G M (1997). Inhibition of a photosystem II electron transfer reaction by the natural product sorgoleone. *Journal of Agriculture and Food Chemistry* 45, 1415-1421.

Green, J. 2000. Adjuvant outlook for pesticides. Pesticide Outlook. October: 196-199.

Green, J.M. and Green, J.H. 1993. Surfactant structure and concentration strongly affect rimsulfuron activity. Weed Technology 7(3): 633-640.

Gupta, P.K., A Hand Book of Soil, Fertilizers and Manures (2004), Agrobios (India), Jodhpur.

Gustafson A.F. Hand Book of Fertilizers (2003), Agrobios (India), Jodhpur.

Hejl A M, Einhellig F A and Rasmussen J A (1993). Effects of juglone on growth, photosynthesis, and respiration. *Journal of Chemical Ecology* 19, 559-568.

Hiran K.S. Jaspal Singh and Acharya M.S. Irrigation Scheduling (1990), C.B.S Publishers and Distributors, New Delhi.

Inderjit and Weiner J (2001). Plant allelochemical interference or soil chemical ecology? *Perspectives in Plant Ecology* 4, 3-12.

Inderjit and Weston L A (2000). Are laboratory bioassays suitable for prediction of field responses? *Journal of Chemical Ecology* 26, 2111-2118.

Inderjit and Weston L A (2003). Root exudation: an overview. *In* Root Ecology; H deKroon, Ed.; Ecological Studies168. Springer-Verlag Press: London, pp. 235-255.

Kato-Noguchi H, Ino T and Sata N (2002). Isolation and identification of a potent allelopathic substance in rice root exudates. *Physiologie Plantarum* 115, 401-405.

Kilronomos J N (2002). Feedback with soil biota contributes to plant rarity and invasiveness in communities. *Nature* 417,67-70.

Kimura Y, Kozawa M, Baba K and Hata K (1983). New constituents of roots of *Polygonum cuspidatum. Journal of Medicinal Plant Research* 48, 164-168.

Larson DL, McDonald S, Fiviz.zani AJ, Newton WE, Hamilton SJ (1998) Effects of the herbicide atrazine on *Ambystoma tigrinum* metamorphosis: duration, larval growth, and hormonal response. Physiological Zoology 71(6):671-679.

Lenka D., Irrigation and Drainage (1991), Kalyani Publishers, Ludhiana.

Macias F A (2002). New approaches in allelopathy, challenge for the new millenium. *Third World Congress Allelopathy Abstracts*, 38.

Masiunas J B, Weston L A and Weller S C (1995). The impact of rye cover crops on weed populations in a tomato cropping system. *Weed Science* 43, 318-323.

Mattice J, Lavy T, Skulman B, and Dilday, R (1998). Searching for allelochemicals in rice that control ducksalad. *In* Allelopathy in Rice: Proceedings of the Workshop on Allelopathy in Rice; M Olofsdotter, ed. International Rice Research Institute, Manila. pp 81-98.

Michel A.M. Irrigation Theory and Practice (1978), Vikas Publishing House Pvt. Ltd., New Delhi.

Misra R.D. and Ahmed M. Manual on Irrigation Agronomy (1987), Oxford and IBH Publishers Co., New Delhi.

Moyer J R, Blackshaw R E, Smith E G and McGinn S M (2000). Cereal cover crops for weed suppression in a summer fallow-wheat cropping sequence. *Canadian Journal of Plant Science* 80, 441-449.

Muir A D and Majak W (1983). Allelopathic potential of diffuse knapweed (*Centaurea diffusa*) extracts. *Canadian Journal of Plant Science* 63, 989-996.

Muller C H (1969). Allelopathy as a factor in ecological process. *Vegetatio*18, 348-357.

Murty V.V.W. Land and Water Management Engineering (1985), Kalyani Publishers, Ludhiana.

Mwaja V N, Masiunas JB, Weston L A. (1995). Effects of fertility on biomass, phytotoxicity, and allelochemical content of cereal rye. *Journal of Chemical Ecology* 21, 81-96.

Nagabhushana G G, Worsham A D and Yenish J P (2001). Allelopathic cover crops to reduce herbicide use in sustainable agricultural systems. *Allelopathy Journal* 8, 133-146.

Nair M, Whitenack C J and Putnam A R (1990). 2,2'-Oxo-1,1'-azobenzene: a microbially transformed allelochemical from 2,3 benzoxazolinone. *Journal of Chemical Ecology* 16, 353-364.

Nimbal C I, Pedersen J F, Yerkes C N, Weston L A and Weller, S C (1996a). Phytotoxicity and distribution of sorgoleone in grain sorghum germplasm. *Journal of Agriculture and Food Chemistry* 44, 1343-1347.

Nimbal C I, Yerkes CN, Weston LA, and Weller SC (1996b). Herbicidal activity and site of action of the natural product sorgoleone. *Pesticide Biochemistry and Physiology* 54, 73-83.

Olofsdotter M, Jensen L B and Courtois B (2002). Improving crop competitive ability using allelopathy - an example from rice. *Plant Breeding* 121, 1-9.

Opik, Helgi; Rolfe, Stephen A.; Willis, Arthur John; Street, Herbert Edward (2005). The physiology of flowering plants (4th ed.). Cambridge University Press. p. 191. ISBN 978-0-521-66251-2.

Pande S.C. Principles and Practices of Water Management (2003), Agrobios (India), Jodhpur.

Parihar S.S. and Sadhu B.S. Irrigation of Field Crop : Principles and Practices, Indian Council of Agricultural Research, New Delhi.

Parmar, B.S. and Tomar, S.S. 2004. Pesticide formulation: Theory and practice. CBS Publishers and Distributors, New Delhi.

Petersen J, Belz R, Walker F and Hurle K (2001). Weed suppression by release of isothiocyanates from turnip-rape mulch. *Agronomy Journal* 93, 37-43.

Putnam A R (1986). Can it be managed to benefit horticulture? *HortScience* 21, 411-413.

Putnam A R (1988). Allelochemicals from plants as herbicides. *Weed Technology* 2, 510-518.

Putnam A R and Duke W O (1974). Biological suppression of weeds: Evidence for allelopathy in accessions of cucumber. *Science* 185, 370-372.

Putnam A R and Duke W O (1978). Allelopathy in agroecosystems. *Annual Reviews of Phytopathology* 16, 431-451.

Putnam A R and Weston L A (1986). Adverse impacts of allelopathy in agricultural systems. *In* The Science of Allelopathy, eds. A. R. Putnam and C. S. Tang. John Wiley and Sons, New York.; pp 43-56.

Qasem J R and Foy C L (2001). Weed allelopathy, its ecological impacts and future prospects: a review. *In* Allelopathy in Agroecosystems,. eds. R. K. Kohli; H. P. Singh and D. R. Batish. Haworth Press, New York. pp 43-119.

Rai M.M. Soil Science, plant chemistry, manures and fertilizers (1985), Ramprasad and Sons, Agra–3.

Reddy S.R. Principles of Agronomy (1999), Kalyani Publishers, Ludhiana.

Rice E L (1984). Allelopathy, Academic Press, Orlando, FL. 422 pp.

Ridenour W M and Callaway R M (2001). The relative importance of allelopathy in interference: the effects of an invasive weed on a native bunchgrass. *Oecologia* 126, 444-450.

Rimando A M and Duke S O (2003). Studies on rice allelochemicals. In *Rice: Origin, History, Technology and Production*; ed. C. W. Smith. John Wiley and Sons, New York. pp 221-244.

Ross MA, Childs DJ (1996) Herbicide Mode-of-Action Summary. Purdue University, Department of Botany: Plant Pathology, West Lafayette IN. Report No. WS-23-W.

Scheffler B E, Duke S O, Dayan F E and Ota E (2001). Crop allelopathy: enhancement through biotechnology. *Recent Advances in Phytochemistry* 35, 257-274.

Sene M, Gallet C and Dore T (2001). Phenolic compounds in a Sahelian sorghum (*Sorghum bicolor*) genotype (ce(145-66)) and associated soils. *Journal of Chemical Ecology* 27, 81-92.

Shepard JP, Creighton J, Duzan H (2004) Forestry herbicides in the United States: an overview. Wildlife Society Bulletin 32(4):1020-1027.

Shilling D G, Liebl R A and Worsham A D (1985). Rye (*Secale cereale* L.) and wheat (*Triticum aestivum* L.) mulch: the suppression of certain broadleaved weeds and the isolation and identification of phytotoxins. *In* The Science of Allelopathy; eds. A. R. Putman and C. S. Tang, John Wiley and Sons Inc: New York, pp 243-271.

Siemens D H, Garner S H, Mitchell-Olds T and Callaway R M (2002). Cost of defense in the context of plant competition: *Brassica rapa* may grow and defend. *Ecology* 83, 505-517.

Singh H P, Batish D R and Kohli R K (2001). Allelopathy in agroecosystems: an overview. *In* Allelopathy in Agroecosystems,eds. R. K. Kohli; H. P. Singh and D. R. Batish. The Haworth Press, New York. pp 1-41.

Sivanappan R.L. and K.R. Karan Gowder. Irrigation and Drainage, Agrobios (India), Jodhpur.

Srivastava, L. M. (2002). Plant growth and development: hormones and environment. Academic Press. p. 140. ISBN 0-12-660570-X.

Swarup R, Perry P, Hagenbeek D *et al.* (July 2007). "Ethylene upregulates auxin biosynthesis in Arabidopsis seedlings to enhance inhibition of root cell elongation". Plant Cell 19 (7): 2186–96. doi:10.1105/tpc.107.052100. PMC 1955695. PMID 17630275. Weier, Thomas Elliot; Rost, Thomas L.;

Tate TM, Spurlock JO, Christian FA (1997) Effect of glyphosate on the development of *Pseudosuccinea columella* snails. Archives of Environmental Contamination and Toxicology 33:286-297.

U.S. EPA (2003) Ambient Aquatic Life Water Quality Criteria for Atrazine: Revised Draft. Office of Water, Office of Science and Technology, Health and Ecological Criteria Division, Washington DC. EPA-822-R-03-023.

Vaughn S F and Berhow M A (1999). Allelochemicals isolated from tissues of the invasive weed garlic mustard (*Alliara petiolata*). *Journal of Chemical Ecology* 25, 2495-2504.

Weier, T. Elliot (1979). Botany: a brief introduction to plant biology. New York: Wiley. pp. 155–170. ISBN 0-471-02114-8.

Weston L A (1990). Cover crop and herbicide influence on row crop seedling establishment in no-tillage culture.*Weed Science* 38:166-171.

Weston L A (1996). Utilization of allelopathy for weed management in agroecosystems. *Agronomy Journal* 88, 860-866.

Weston L A and S O Duke (2003). Weed and crop allelopathy. *Critical Reviews in Plant Sciences* 22: 367-389.

Weston L A, Harmon R and Mueller S (1989). Allelopathic potential of sorghum-sudangrass hybrid (sudex). *Journal of Chemical Ecology* 15, 1855-1865.

Weston L A, Yang X and Scheffler B E (2002). Gene regulation by bioactive root exudates produced by *Sorghum* spp. *Third World Congress Allelopathy Abstracts* 76.

Willis R J (1997). The history of allelopathy. 2. The second phase (1900-1920). The era of S. U. Pickering and the USDA Bureau of Soils. *Allelopathy Journal* 4, 7-56.

Willis R J (2000). *Juglans* spp., juglone and allelopathy. *Allelopathy Journal* 7, 1-55.

Wink M (1999). Introduction: biochemistry, role and biotechnology of secondary metabolites. *In* Functions of Plant Secondary Metabolites and Their Exploitation in Biotechnology, ed. M. Wink. CRC Press, Boca Raton FL. *Annual Plant Reviews, Volume 3*; pp 1-16.

Index